THE POWER OF NUMBERS

Stilovsky
and
Schrödinger

authorHOUSE

AuthorHouse™ UK
1663 Liberty Drive
Bloomington, IN 47403 USA
www.authorhouse.co.uk
Phone: 0800.197.4150

© 2018 Stilovsky and Schrödinger. All rights reserved.

Initial cover design by John Hocknull
By the same authors: Hoggrills End (published Dec 2017)

No part of this book may be reproduced, stored in a retrieval system, or transmitted by any means without the written permission of the author.

Published by AuthorHouse 02/09/2018

ISBN: 978-1-5462-8818-3 (sc)
ISBN: 978-1-5462-8817-6 (e)

Print information available on the last page.

This book is printed on acid-free paper.

Because of the dynamic nature of the Internet, any web addresses or links contained in this book may have changed since publication and may no longer be valid. The views expressed in this work are solely those of the author and do not necessarily reflect the views of the publisher, and the publisher hereby disclaims any responsibility for them.

FOREWORD

In today's world the use of numbers grows by the day and we depend on them for so much. This book contains a series of lists which contain information about numbers and their use in society. They will be most useful to those of a *quizzical* nature but should be of general interest to all.

Readers should remember that lists like these are never complete and it's possible that the odd error may have crept in so if you propose to use any of the information contained herein, please remember that it is your own responsibility to verify it before such use.

If you notice any errors we would be most grateful if you would notify them to us at our website which can be found at: https://felixschrodinger.wordpress.com/

Pyotr and Felix

CONTENTS

"Numbers"		vi
Introduction	A short story on numbers	vii
Numbers	Their use in everyday life	1
Ordinals	First, second, third, etc…	38
Groups	Miscellany	42
American Aircraft	Their names and numbers	90
Anniversaries	Never to be forgotten	95
Atomic Numbers	And 'magic numbers'	97
Bingo Numbers	Top of the Shop!	102
Comparatives	Ascii code and bases	106
Constants	Life of Pi and others	111
Containers	How big is your pot?	113
Definitions	Meanings	115
Education Maths	When to teach	124
E Numbers	They're in everything	126
Foreign Numbers	En Francais et Espanol etc…	135
Frequency	Time periods	137
Games and Sport	Teams	140
Gas Marks	And isotopes	145
Geometry	An angle on it	146
Geo Coordinates	Lat and Longitude	148
The Seven Natural Factors of Greek Medicine		150
Heralds	How many of each?	152
Indexes	Stock Exchanges	154
Music and Sound	How loud does it get?	156
Paper and Books	What's in a quire?	158
Rhymes and Song	Four and twenty black birds..	161
Roman Numerals	MDCLXVI (1666)	177
Scales and grades	How strong is it?	178
Units	Where we start from	198
Vehicle Reg's	UK car registrations	202
Dial Codes	International Dialling Codes	206
Strange Numbers	Other Numeric systems	211
Appendix A	Songs with numbers in the title	213
Appendix B	Films with numbers in the title	221
Appendix C	Books with numbers in their titles	241
Appendix D	Numbers which occur in the Bible	249

"NUMBERS"

A 'number' may be one of many things, and there are numerous dictionary definitions to contend with which include both nouns and verbs. Its most obvious context is that of an arithmetical value such as: one, ten or a hundred which are commonly used in counting and in making calculations. It may also be a numerical value indicating position in a list, series or even a location; sometimes for identification. Another meaning indicates a host (or even a hoard) of something and others mean the issue of a magazine or a musical composition. There's even a book of the Bible devoted to them and many everyday expressions refer to them:

A large computer may be referred to as a *number cruncher* and the first mate on a ship is *number one*. Vehicles have a *number plate* for identification and an activity involving chance is *a numbers game*. The UK's PM resides at *Number 10* and an assistant is often called their *Number Two*. If you are at risk from ordinance then a bullet *may have your number on it* and then *your days are numbered. What's your number* is a key way of making contact but *I've got your number* has a completely different connotation.

She was wearing a little black number indicates style but was she also *adept at numbers*? He was *numbered as one of their friends* but was he amongst those *numbered in the thousands as any number of them* could have been. You could read this *in a number of ways* or, alternatively you could *just do it by the numbers*. They appeared *to be without number* and *did a number on him* so I must go – *my number just came up*!

INTRODUCTION

I called the dentist's for an urgent appointment.

"Tooth hurty today?" I hear the receptionist say and assume she is from Hong Kong. "Of course," I reply, "that's why I rang."

"That's confirmed 2.30 with Dr Frost." She says, "What name is it?"

I'm told that dentist's receptionists have this interchange included in their basic training these days and are required to laugh when you make a joke of it. But why are numbers so important to us. Well this sort of interchange demonstrated one aspect – how much time depends on numbers. It required the delivery and acceptance of a number to indicate when a meeting should take place and that was numeric.

Actually time is somewhat unusual as most of the numbers we use are based on ten. This is called the *base* and no-one actually knows why we use ten though it is widely assumed that it is because we have ten fingers and toes. Why not twelve which would be more useful when filling egg boxes? Base ten is generally known as the *decimal system* and is widely used by mathematicians and scientists as it is the most accepted system. Having said that, it's not the only one.

Time, as mentioned above is more complex. Short-term time uses base 60 i.e. we have sixty seconds in a minute and sixty minutes in an hour. However it then goes completely astray (for good reason) as we have 24 hours in a day and then 365¼ days in a year. Obviously the latter larger units relate to the rotation of the Earth and its path around the Sun; with some complication as regards to the moon which has a base of nearly thirty but not quite. We handle this by ignoring the arithmetic and handing out arbitrary names for the months which is why the religions of the world spend so much time arguing about the dates on which festivals fall.

When we get to larger units of distance we revert to decimal and talk about tens, hundreds, thousands and millions of kilometres or miles. When these become inadequate we bring in the speed of light and use light years; again in decimal terms. Getting smaller than seconds involves decimals going right down to the *nano* level and beyond. It's also convenient, in the decimal system to use powers thus if I add a little number up above and to the right of another number that raises it to the power of the little number. For very big numbers this is very useful and avoids having to write a lot of zeros.

So time involves a complex set of bases but what do computers use? A much simpler system altogether called *binary* which basically means two. This is easier for electronic machines to understand as it only uses '0' and '1'. It's very easy to design circuits and switches which just use these two digits rather than more complex numbers – that's until you get human intervention. Endless strings of 1s and 0s would be difficult to work with so an intermediate base was brought in - *octal*. This uses eight digits at a time and was useful for many years until it was found necessary to introduce *hexadecimal* or base 16. This has numbers 0, 1, 2, 3, 4, 5, 6, 7, 8, 9, A, B, C, D, E, F and G to be able to work up to sixteen with single digits. Soon after, 32 digits were introduced and some now use 64 as a base.

Was twelve ever used? – well yes – before decimalization in the UK we had twelve pennies to a shilling and we still have twelve inches to a foot. So what about twenty? – twenty shillings to a pound. Thank goodness we got rid of some of them. Engineers in the UK converted to decimal coinage in the late 1960s and to decimal measurement in the early 1970s. Prior to that a 'bill of quantities' was measured in yards, feet and inches (and even eighths of an inch) with unit costs in pounds, shillings and pence. A penny was one 240th of a pound.

Time was not the only system to use sixty as a base; it was, and still is used for latitude and longitude though, of late, decimal degrees tend to be used rather than degrees, minutes and seconds. In spite of all the progress made towards decimalization, the speed of ships is still rated in knots or *nautical miles* – a measure based on latitude and longitude. These systems, using four dimensions, enable us to define a place in three dimensions and time so that meetings can take place; without them life would be quite difficult.

There are other systems which mix numbers with letters which appear to be numeric but are actually nominal – i.e. they are names. The M6, A303 and B4098 are names rather than numbers. The test for this is whether you can add or subtract them from one another and obviously you can't subtract the A30 from the A303. Zip codes and postcodes, similarly as they have no numeric significance. They indicate places which need interpretation to give them meaning. House numbering is somewhere in between a name and a number. They are obviously names as they identify particular properties but they also have some aspect of a number as they increase or decrease as you progress along a road. Are they true numbers – probably not - as adding or subtracting them makes no sense.

GPS coordinates have become very useful now that so many satellites circle the Earth though in apparently stationary orbit. These use lots of complex calculations to let you know your precise location on the surface of the planet and how to get from one place to another using, initially names but actually, underlying coordinates which are large numbers. Incorporated into such systems are complex equations using geometry, trigonometry and algebra which are the second order mathematical tools after addition, subtraction, multiplication and division. Navigation, engineering and architecture all require a clear understanding and skill in using them.

Algebra, a mystery to many, is a rather simple way of using letters in place of numbers when you want to be able to vary the numbers (or parameters as they are called) before you get to an answer.

Calculating anything more complex than the very simple requires equations which spell out the relationships between things without having to assign numbers. All of the sciences depend heavily on the use of equations, especially physics.

Possibly the most confusing numbers are statistics. These are numbers which represent such things as samples of a large number of things – a *population*. If I say that 6% of the population is unemployed then most people will know what I mean and if I say that it fell by 1% they will also be comfortable. But if the average rise in house prices rose by 10% would it be 10% of the average price or 10% of the previous price rise? Clarity is all when dealing with statistics; as they say: "There are lies, damned lies and statistics."

Did you remember that the telephone system, when first introduced, just used names and only in the 1970s did it go wholly onto numbers. The early telephone dials has three letters with each number and the exchanges all had names which eventually became numbers. Can you imagine not using numbers now?

Photography used to be a wet process using chemicals and hardly anyone cared about the grain in a film as it was so fine. They did however, after about 1970, care about the film speed which was an ISO number and the higher the number, the faster the film responded to light. When digital cameras came into use, we forgot about film speed and got fixated on pixels and aspect ratios, all of which come in number format. This affected file size and your ability to save a picture or transmit it elsewhere.

Which brings us on to the internet. Every computer has an IP – a unique number which identifies it and actually it's not a number – it's a numeric name. But the speed of our internet connection is definitely a number measured in *baud*. Don't ask why.

Where would sport be without numbers? Generally there are two kinds of sports – those which use scoring and those that involve winning. You don't actually need numbers in events which involve getting past a winning line before other competitors but we still have first, second and third for medals. After that the times and heights are measured and form endless tables of records – all numerical. But some forms of athletics use distances as the measure of success; the throwing events being the best examples and no-one knows who the winner is until the last throw. Jumping is somewhat similar.

Sports which involve low scores use simple numbers. These include soccer, field hockey, ice hockey and others which use 'goals' as a means of differentiation. The winners get points, usually three for a win and one for a draw and then numerical tables are compiled – league tables. Baseball is somewhat similar in its scoring though lacking goals. High scoring games such as cricket have all sorts of complications and rules to comply with; even a formula to determine who has won in the event that a match is interrupted by rain. In between these extremes are medium scoring games such as Rugby, American Football and Aussie Rules whereby there are a number of ways that you can score points.

Games have similarities. Snooker and Go involve relatively big numbers whereas chess and most board games involve clear winners and losers. Monopoly is more complex as it can go either way depending on whether the players have time to complete the game or agree to finish at a certain time. In the first case all players except the winner are bankrupted and, in the latter case, there is a numerical valuation of assets to determine who has won.

Talking of games brings us on to gambling which has its own version of mathematics to support what has become a major business. Odds are the backbone of any sort of gambling where the wager is more complex than a win or lose bet. Any bookmaker with a poor understanding of odds would soon go bankrupt.

Most card games use values rather than numbers which is evidenced by the face cards being known as members of a royal family rather than being assigned numbers directly. This is further confused by the Ace which, though the lowest numerically, is the highest ranked. Pontoon requires gamblers to get as close to the number 21 without exceeding it and has many popular variations including Blackjack.

Dice are not only used in gambling (craps) but are prevalent in maths as well when we start to consider the chances of events happening. What is the chance of throwing a six and what is the chance of a coin coming up heads six times in a row? How often will a *one in a hundred years* flood occur? And, what is the chance of another one next year if one occurred last year?

How would business survive without numbers? Accounts have to be compiled and annual reports produced. Whilst the Chairman's words may be read, it's the numbers which count. Spreadsheets proliferate and where would we be without them? Wages and salaries are paid out based on the local currency and relate back to a numerical assessment of the workers input. Share prices are avidly followed and we even have two major television channels which contain nothing else. The price of your local currency in relation to the dollar is of great importance and is reported by the minute to a fraction of 1%.

Grading systems often use numbers. Typical of these are the Beaufort Scale for wind speed and Moh's system of hardness grading; the latter of which is based on the numbers one to ten. More commonly, these days are systems which grade anything involving performance and condition in simple grades one to five. Exam results are usually reported in grades but their base is nearly always a percentage of marks in relation a maximum of 100%.

Even my health depends, to some extent, on numbers. The prescription for my glasses is numeric and my hearing test is reported both graphically and numerically. My height is 177cm (5'10") and my weight exactly 70kg (11 stone or 154 lbs) which makes my BMI (Body Mass Index) 22.3.

"Excuse me, Mr Sikorsky, can you spare me a few minutes," asks the pharmacist, "I want to review your medication?" I correct her over my name which is Stilovsky and we sit down at the desk in the little interview room and she looks at a printout in front of her. "Let's start with the bisoprolol; are you on the 2.5 or the 5.0?" "only 1.25," I respond, "even the 2.5 made me drowsy and ruined my sex life." "Ok, what about the aspirin – 75 gastro-resistant?" "Ok," I said, "No problem." "Tamulosin 400 – one or two a day?" "Just the one and they have made a considerable improvement – I only pee once in the night since taking them." "And the lanzoprazole 15 – one or two a day?" "No, just one when I expect to have a problem like when I eat curry." "And you recently started on the 2mg perindopril – how's that?"

I had just come back from the Doc's. She checked my blood pressure and said 120 over 60, pulse rate 63; "That's good," I responded, "I'm normally 135 over 65 and my pulse is usually 60 or below."

I drove back home in my 2010 Insignia 2000 CRX with the 1956cc supercharged diesel engine which pushes out 138bhp at 2500rpm. It's one of fifty shades of boring grey and is hardly a Bugatti Beyron (or is it Veyron?) as it does 0-60 in a quite pedestrian 10.5 seconds. However it does record 58mpg if I drive carefully (4.8 litres/100km) but don't check that as I made it up). Jeremy Clarkson would only ever look at it to tell you it's not even a Vauxhall but an Opel import made near Frankfurt but then he's too busy railing against the 50mph speed limit in motorway roadworks. When I get home I wonder whether to have a beer (5% abv and 1.8 units) or a glass of wine (12% and 1.5 units). Neither will put me over my weekly limit of 25 units and I am not driving again today so will not run any risk of being over 60 or 70mg per whatever it is.

Page numbers are commonly used in books and could be viewed as names in the same way that house numbers are viewed, however, as they come in strict order they should be viewed as numbers. Complex systems are also derived to label books such as ISBN which are actually numerical names. Dewey is another system which is used by libraries to decide on the shelf placing of a book.

Numbers are essential to the way we live in modern society though one could not determine whether they are more or less important than the written or spoken word? We simply could not survive in the world today without them which makes it essential that we protect their use and clamp down on their misuse.

Anyway I have to break off as I have a severe pain in my chest (grade seven on a scale of one to ten) and have to call 999 (911 for some of you and 112 on your mobile) to request an ambulance. Before that I need to take a 300mg aspirin. I'm told that time is all important and 87% of all suspected heart attack emergencies are successfully resolved. I do hope, however, that I am one of the 97% of those who are responded to within the eight minutes that the emergency ambulance is supposed to take getting to me and that I am seen within the 96% target of admissions that are seen within four hours when admitted.

[They say that 75% of all statistics are made up on the spot and I just made most of these up – except for the telephone numbers that is.]

PS - just got home in time for a *number two* before I watch *NUMB3RS* on the telly.

PPS - Here are some of Felix' essential numbers in case you want to steal his identity:

Pet Passport (chip): 999111666
National Insurance: XX 99 88 55X
Driving licence: SCHRO606099FEUK 99
Credit card: 4444 1111 2222 3333
Debit card: 9999 8888 7777 6666
Bus pass: 666555 0000 1111 4444
Coop dividend card: 55555
Pets 'R' Us Loyalty card: 666555 77 2222 3333

Pyotr

NUMBERS

0

Zero	None
Nought	Nil
Oh	Wipe-out
Cipher	Zilch
Shutout	Duck
Clean sheet	Blank
'00' = licence to kill	Nix
Maiden over in cricket	Nothing
Zero hour	Love
Not used in Roman numerals	Misere in Solo

1

Ace	Unit(y)
First square number	First cube number
Specific gravity of water	Atomic weight of hydrogen
One in the eye	Atomic number of hydrogen
Wheels on a unicycle	Hump on an Arabian camel
Solo (alone)	Snooker red ball value
Cell in an Amoeba	One Nation
One of these days.....	One of those days (things)
One foot in the grave	One .... short of a ......
One hit wonder	One over the eight
One small step for man	One giant leap for mankind
One for the pot	One for the road
One night stand	One armed bandit

One horse town
One man band
One stop shop
One off the wrist
One eyed trouser snake
One-step (dance)
One-parent family
One too many for me
One-eyed Jacks
No 1 - *Thomas the Tank Engine*
Go one better
Just One Cornetto
Put one on you (or over you)
Number One (1st Lt. in Navy)
Mach one - the speed of sound
One flew over the cuckoo's nest
One swallow does not make a summer
One a penny, two a penny, hot cross buns

One track mind
One-upmanship
One for his Nob in crib
One man and his dog
One degree under
The one that got away
One-liner
One and all
One Above (God)
All in one
Hole in one
At one fell swoop

2

Deuce
Couple
Duo
First even number
Pints in a Quart
Teaspoons in a dessertspoon
Number two (defecation)
Cupfuls in a pint
Stirling Moss
Humps on a Bactrian Camel

Pair
Twin
Duet
Binary base
Gallons in a peck
Tablets of stone (Moses)
Snooker yellow ball value
Oars for a sculler
2LO - London's 1st call sign
Strings to one's bow

Wings on a biplane
Croquet team
Two hairs past a freckle
Two lovely black eyes
Two by two
Two for tea
Two heads are better than one
Two Way Family Favourites
Two peas in a pod
Two Eyes of Greece
Two's company……
Two-bit
Two-step (dance)
It takes two to tango
Thick as two short planks
Dessertspoon in a tablespoon
Two Ronnies (Barker and Corbett)
Conversion points (Rugby Union)
A volunteer is worth two pressed men
Two eyes of Greece (Athens & Sparta)
Two penny tube (London Central Line)
A bird in hand is worth two in the bush
Persons to be joined in Holy Matrimony
Two fishes (+5 loaves) to feed the 5,000
One a penny, two a penny, hot cross buns

Two bits = a quarter
Cannon score (billiards)
Two minute silence
Two tone horn
Two's company
Two Gentlemen of Verona
Two ton Tessie (O'Shea)
For two pins
Two penny Opera
Two Sicilies
Two-Tone (record label)
Tale of Two Cities (Dickens)
The Tamworth Two (Pigs)
Put two and two together

3

Trey
Prial (cards)
Feet in a yard

Triad
Trinity
Miles in a league

Cardinal virtues
Kingdoms in nature
Peaks on a tricorn hat
Primary colours
Scruples in a dram
Colours on a traffic light
Wheels on a Reliant Robin
Minutes in a round at boxing
'Eyes' on a coconut
Strikes in a turn at conkers
Green ball value (snooker)
Sides of a triangle
Legs on a trivet
Gods in tri-theism
Men in a Triumvir
Legs on a Triskelion
Stripes in a tricolour
Trigon (Astrology)
Three Men in a Boat
Three cheers
Three wishes
Three-ply
The Three Graces
Three Furies
Three Virtues
The Three Stooges
Three phase electricity
Three sheets in the wind
Three of a kind
Three Counties Show

Legs on a Tripod
Inalienable rights
Heads of Cerberus
Secondary colours
Wheels on a tricycle (trike)
'3 D' (or dimensional)
Norns (Norse mythology)
Three ring circus
Balls in billiards
Ossicles in the ear
Strikes and out in baseball
Wings on a triplane
Books in a trilogy
Pictures in a triptych
Subjects in a Tripos
Lobes on a trefoil
Spikes on a trident
Three Blind Mice
Three men on the Bummell
Three point turn
Three-legged race
Three 'R's
Three Fates
Three Harpies
Three little pigs
Three coins in the fountain
Three star general
Three's a crowd
The Three Musketeers
Three Degrees (group)

Three Mile Island
Three-day eventing
Three Kings' Day (12th Day)
Rule of three
Three cornered fight (election)
Three line whip (parliament)
Three Choirs Festival
Three field system (crop rotation)
Three decker (warship)
Witches on the Blasted Heath
Triple thread of Brahmanism
Three tailors of Tooley Street
Holes in a ten pin bowling ball
League of the Three Emperors
Goldilocks and the Three Bears
Three legged mare (the gallows)
Two's company, three's a crowd
Three acres and a cow (Collings)
Christ denied three times by Peter
Three times three (Hip Hip Hooray)
Three Wise Men (or Kingsoforientar)
Ménage a trois (three in the kitchen)
Three mile limit (was territorial waters)
Little maids from school (*The Mikado*)
Battle of the Three Emperors (Austerlitz)
Three pegs for the Tower of Hanoi (puzzle)
Golden balls of Lombardy (hence pawnbrokers)
The Adventure of the Three Gables (Sherlock Holmes)
The Adventure of the Three Students (Sherlock Holmes)
The Adventure of the Three Garridebs (Sherlock Holmes)

Three bags full
Three-day measles (rubella)
Three Estates
Threesome
Three-gaited (horse)
The three F's
Three Gorgons
Three tongues

4

Rods in a chain
Pecks in a bushel
Quarters in a Cwt
Roods in an acre
Firkins in a barrel
Inches in a hand (horses)
Cardinal points on a compass
Gospels in the New Testament
Grand Slam tennis tournaments
A boundary in cricket
Seasons in the year
Faces on a tetrahedron
Stomachs for most ruminants
Number in a curling team
Kings of a French pack
Four-ball (golf)
Snooker brown ball value
Four Just Men (Edgar Wallace)
Four Horsemen of the Apocalypse
Four in hand
Gang of Four (Social Democrats)
Four leaf clover
Four letter word
Four rivers of Paradise
Four of a kind (poker)
Four Pennies (group)
Four Tops (group)
Four colour problem (maths)
Four corners of the Earth

Nips in a bottle
Quarts in a Gallon
Balls to walk in baseball
Gills in a pint
Hogsheads in a tun
Quadrants in a circle
Quarter days
Beatles
Balls in Croquet
Suits in a pack of cards
Sides in a quadrilateral
Laps in a speedway heat
Number in a polo team
Tastes (actually now five)
Bases on a baseball field
Tetrad (4 letters e.g. JHVH)
Four letter (****) man
Foursquare
Four in the morning
On all fours
Annals of the four masters
Four poster bed
Four stroke (engine)
Unit Four plus Two (group)
Four star petrol
Four Seasons (the group)
Four ale bar (the cheapest)
Four eyes (wearer of specs)
Four minute mile

Four freedoms (FD Roosevelt)
Foursome
Four-plus cover (advertising)
Four colour process (printing)
Corners and sides on a Square
Chambers in the human heart
Old Four legs (the Coelacanth)
Strings on a violin (and a base guitar)
The Sign of the Four (Sherlock Holmes)
Four minute men (hence "minuteman")
Four Marys (Mary Stewart's companions)
Grand Slam = 4 runs on one hit in baseball

Four last things
Four-leaf clover

Four ale (4d a pint)

5

Years (etc) in a Pentad
ml (or cc) in a teaspoon
Snooker blue ball value
K's of Sikhism
Points of Calvinism
Five pounds = Lady Godiva
Points for a Try
Vowels in the alphabet
Senses
Tentacles on a starfish
Points on a pentangle
Cents in a nickel
Classics in horseracing
Number in a basketball team
Five pins (Canadian game)
Five-a-side football

Events in a pentathlon
Carats in a gram
Lines in a Limerick
Olympic Rings
Wits (Shakespeare)
Five-spot (5 dollar bill)
Fiver
Great Lakes
Pillars of Islam
D Day beaches
Sides to a pentagon
Books in the Pentateuch
Victims of Jack the Ripper
Five year plan (USSR)
Fives (the game)
Five-o'clock shadow

- Five finger exercise
- Five-stones (jacks)
- Five mile high club
- Five guys named Moe
- Five card trick
- Five little pigs went to market
- Five alive (Short Circuit)
- Take five
- *The Famous Five* (Enid Blyton)
- "I'll give it foive" (Janice Nicholls)
- Radio Five Five Live
- Five Nations (American Indians)
- Five-faced bishop (moschatel)
- Members of the Long Parliament
- The five towns (Arnold Bennett)
- Minutes to look for a lost golf ball
- Five Classic Orders of Architecture
- Coins per player in shove ha'penny
- *The Five Orange Pips* (Sherlock Holmes)
- Five loaves (+ two fishes) to feed the 5,000
- Permanent Members of the Security Council
- Seconds for the tape to destruct in Mission Impossible
- Five Nations (Kipling)
- Five boys chocolate (Fry's)
- Five Star (hotel and group)
- Five year plan
- Fireball XL5
- Dave Clark Five
- Bunch of fives
- Jackson Five
- Five Mile Act
- Full fathom five... (Tempest)

6

- First perfect number
- Balls in an over (cricket)
- Six-pack
- Boundary (without bouncing)
- Sides in a hexagon
- Faces on a dice
- Counties of Northern Ireland
- Feet in a fathom
- Musicians in a sextet
- Games to win a set in tennis
- Least number of darts for 301
- Coloured balls in snooker
- Members of a volleyball team
- Members of an ice hockey team
- Brownie guide division
- Sixer = Cub leader
- Archangels of the Bible
- Six Five Special
- Six Day War (1967)
- 'Les Six' - French composers
- Six stringed whip
- Six Nations (American Indians)
- Six hooped pot (2 quarts)
- Days to make heaven and earth
- Touchdown in American Football
- Six foot way (between railway tracks)
- Pockets on a snooker (or pool) table
- Six of one and half a dozen of the other
- The Six Clerks' Office (the Court of Chancery)
- Firkins in a hogshead
- Wives of Henry VIII
- Strings on a guitar
- Pence in a 'tanner'
- Faces to a cube
- Insects' legs
- States of Australia
- Points on the Star of David
- Pips in the GMT signal
- Greyhounds in a race (UK)
- Six shooter
- Characters in Cluedo
- Weapons in Cluedo
- Stumps in a game of cricket
- Cub scout division
- Snooker pink ball value
- At sixes and sevens
- Six of the best (caned)
- Number Six (*The Prisoner*)
- Coach and six
- Tom Mix (at darts)
- Knock someone for six
- The Six Articles

'The Six' - original members of the Common Market
The Adventure of the Six Napoleons (Sherlock Holmes)

7

Ages of man (Shakespeare)
Hills of Rome (and Sheffield)
Days in a week
Wonders of the Ancient World
Sides in a heptagon
Snooker black ball value
'Sides' on a fifty pence piece
Joys (and sorrows) of Mary
Sacraments
Spiritual works of mercy
Snooker ball colours (excl. white)
Members of a handball team
Fathers of the Church
Seven Sages
Seven Virtues
Seven Samurai
Seven Sleepers (Biblical)
Seven Year Itch
The Magnificent Seven
The 'Seven' (original EFTA)
The Temperance Seven (group)
The Seven Deacons
Members of a water polo team
'Sides' on a twenty pence piece
Red stripes on the American flag
Dance of the seven veils (Salome)

Colours in a rainbow
Deadly sins
Dwarfs
Continents
Champions of Christendom
Corporal works of mercy
Days of creation
Seven Seas
Branches on a Menorah
Wise Masters
Members of a netball team
Members of a Kabaddi team
At sixes and sevens
Liberal Arts
Seven Years War (Germany)
Seven Seas of Rye (Queen)
Seven against Thebes
Seven Sisters (Pleiades)
Blake's Seven
Seven Dials (Holborn)
Seveners (the Isma'ilis)

Years bad luck for a broken mirror
Seven wonders of the Middle Ages
Seven Sisters (chalk cliffs in Sussex)
Seven Weeks War (Austria/Prussia)
Knock seven bells (out of someone)
Seven plagues of Angels (Revelations)
Island of Seven Cities (mythical – Spain)
Seven loaves (+ a few fishes) to feed the 4,000
Seven Baskets used to collect food for the 4,000
Sum of the spots on opposite sides of a dice
Gold medals for Mark Spitz at Munich (1972)
Seven segment display (electronic number display)
Seven maids with seven mops (*Alice Through the Looking Glass*)

8

Octal base
Stone in a hundredweight
Furlongs in a mile
Pints in a gallon
Members in a tug of war team
Octopus' tentacles
Bits in a byte (8 bit)
Sides to an octagon
Corners on a cube
Points on a Maltese Cross
People on Noah's Ark
Pawns each in chess
'arry Tate
Behind the eight ball
One over the eight

Spans in a fathom
Drachms in an ounce
Notes (tones) in an octave
Bushels in a quarter
Quarts in a peck
Spider's legs
Faces on an octahedron
Baby's in a bottle
Rowers in a boat race crew
Greyhounds in a race (US)
8mm cine
Cord feet in a cord (timber)
Eight ball (pool)
Eight to the bar (boogie)

Butterfield Eight (Elizabeth Taylor)
Lanes in an Olympic swimming pool
'Pieces of eight' (parrot in *Treasure Island*)

9

Square feet in a square yard
Look nine ways (squint)
Heads on a Hydra
Skittles
Orders of Angels
Nice as nine pence
Holes in bar billiards
Innings in a baseball match
Points for game in squash
Members of a baseball team
Members of a softball team
Worthies
Queer as a nine bob note
Cat o' nine tails (nine tail bruiser)
Nine dart finish (for 501)
Nine yard sari
Nimble as nine pence
Nine pence to a shilling (simple)
Nine Lives
Inches in a span (not precise)
Novena (nine days of prayer)
Inches wide for cricket stumps
Nine Tailors (maketh the man)
Members of a rounders team
Star Trek: Deep Space Nine (Series)

Gallons in a firkin
Sides in a nonagon
Months of pregnancy
Rivers of Hell
Muses
Rooms in Cluedo
Beethoven symphonies
Times out of ten
Types of crown (heraldry)
Nine Carat gold
Members of a Kho Kho team
Worthies of London
9 mm Browning
Ninepins
The whole nine yards
Dressed up to the nines
Right as nine pence
Marks of Cadency (Heraldry)
Nine Men's Morris (game)

Nine Days Queen (Lady Jane Gray)
Planets in the Solar System (now 8)
Nine points of the law (possession)
Nones - the ninth day before the ides
Highest card in a 'Yarborough' (Bridge)
Nine of Diamonds - the curse of Scotland

10

Base for arithmetic
Chains in a furlong
Cables in a nautical mile (UK)
Legs on a lobster, crab, shrimp
Degrees of hardness (Mohs)
Sides in a decagon
Fingers and toes
Commandments
Cards per player in Gin
10, Downing Street
Hurdles in the 100m and 400m
10 cc (group)
Ten pin bowling
Ten Little Indians
Top Ten
Tenner (£10)
Ten to One (against or on)
The Upper Ten (the aristocracy)
Digits in an ISBN (prior to 2007)
Members of a mens' lacrosse team
The Council of Ten (Venetian Republic)
The 'perfect number' (according to Pythagoras)

Decimal base
Years in a decade
Square chains in an acre
10! (Bo Derek)
Avatars of Vishnu
Faces to a decahedron
Events in a decathlon
Plagues of Egypt
Pins for a strike
Cents in a dime
Mr 10%
Ten green bottles
Ten gallon hat
Ten-four (*Highway Patrol*)
Ten penny nails
Hollywood Ten

Years for the maximum US Presidency (except Roosevelt)
Amendments to the American Constitution in The Bill of Rights (initially)

11

Cricket or soccer team
Days lost from calendar in 1752
Years of tyranny under Charles I
Eleventh chord (jazz)
Eleven Plus for grammar school
Members of a field hockey team
Number of Reg's bus (*On the buses*)
Members of an American Football team

First eleven, second etc...
Unionist states
Confederate states

12

A Dozen
Inches in a foot
Months in a year
Sides in a dodecagon
Signs of the Zodiac
Apostles or Disciples
Provinces of Canada
Pipers (Whisky)
Sides on a threepenny bit
The Dirty Dozen
12" (record, usually remixes)
Members of a shinty team
Baskets used to feed the 5,000
'Twelve pitch' - typewriter typesize
The Twelve Tables (of Roman Law)

Pennies in a shilling ('bob')
Notes in a chromatic scale
Tribes of Israel
Faces on a dodecahedron
Members of a jury
Labours of Hercules
Good men and true
Tables of Roman Law
Twelve tone music
Draughts per player
12 mile limit

The Twelve-step Program of Alcoholics Anonymous
Members of a women's lacrosse team
Members of a Canadian Football team
Wind forces of the Beaufort Scale (up to 17 in the US)

13

A baker's dozen
Players in a rugby league team
Players on a cricket pitch
Lunar months in a Year
Digits in an ISBN (since 2007)
Years of Prohibition in the USA
Stripes on the US flag (original states)
Valkyries
Diners at the Last Supper
Dick Turpin (at darts)
Unlucky for some (bingo)
Principles of (Jewish) Faith

14

Books of the Apocrypha
Pounds in a stone
Operettas of Gilbert & Sullivan
Points of Woodrow Wilson (WWII)
Maximum number of clubs allowed for a pro golfer
Lines in a sonnet
Stations of the Cross
Carbon 14 for dating

15

To score at crib
Men on a dead man's chest
Reds in snooker
First point in tennis
Members of a Gaelic Football team
Fifteen pieces each in Backgammon
'The Fifteen' (first Jacobite rising in 1715)
Rugby Union team
Game in badminton
Members of a hurling team

16

Hexadecimal base
Ounces in a pound
Sweet Sixteen (Billy Idol)
Tablespoonfuls in a cup
Square rods in a square chain
Sweet Little Sixteen (Chuck Berry)
Cubic feet in a cord foot (timber)
Sixteen-string Jack (John Rann, a highwayman)
Sweet sixteen and never been kissed (bingo)

Nails in a yard
Sixteen yard box (soccer)
Drams in an ounce
Quavers in a breve

17

B17 the Flying Fortress
17-year locust

18

Inches in a cubit (approx.)
Holes on a golf course
18" = length of a Plimsoll Line
Members of an Australian rules football team

Age of majority
Carat gold

19

Nineteen to the dozen
Impossible to score at crib
Points to each side of a Go board
Cherubs on a Trivial Pursuit board

Nineteenth hole (the bar)
Years in a metonic cycle

20

Fluid ounces in a pint
20/20 vision
20th Century Fox

Shillings in a pound
Hundredweight in a ton
Grams in a scruple

Heats in a speedway meeting
Troy ounces in a pound
Years asleep for Rip van Winkle
Years in a 'Preston Guild'
Teams in the Premier League
Lovesick maidens (Patience by G&S)
Twenty Questions (radio programme)

Top twenty
Faces on an icosahedron
Quires in a ream
Twenty 20 cricket

21
Gun salute
Pontoon
Vingt et un
Coming of age

Shillings in a guinea
Consonants in the alphabet
Blackjack
Spots on a dice

22
Yards in a cricket pitch
Catch 22
Yards in a chain
Dinkey doo (Housy Housy)
Trumps in the Tarot 'Major Arcana'
Balls on the table at the start of a game of snooker

Snooker balls
Two little ducks
Carat gold

23
Pairs of Chromosomes in a human cell

24
Hours in a day
Pure gold (24 ct)
Grains in a pennyweight
Pompey 'ore (Housy Housy)

Sheets in a quire (now 25)
Pieces in a Chess set
Ribs in the human body
Draughts on a draughtboard

Blackbirds (actually 4 and 20)
Points on a backgammon board
Twenty Four Hours from Tulsa
(Gene Pitney)

25

Silver Jubilee/Wedding	A 'pony' = 25 pounds
Maximum points at Croquet	Points for gin
Outer bull at darts	Sheets in a quire (was 24)
Cents in a quarter	Links in a rod

26

Letters in the alphabet
Miles in a marathon (plus a bit)
'Bed and breakfast' (5+20+1 at darts)
Counties of the Republic of Ireland

27

Three cubed
Cubic feet in a cubic yard
Books of the New Testament (AV)
Bridges over the Thames (reputedly)

28

Pounds in a quarter	Dominoes in a set
Hurdles in a steeplechase	Days in February (normally)
Inches high for cricket stumps	

29

Days in February in a leap year   Maximum in cribbage

30
Pieces of silver
Second point in tennis
Thirty Years War (1618-1648)
Seconds of the *Countdown* clock
Journalists' indication of 'the end'
Thirty Tyrants (Magistrates in Athens)
Fences to jump in the Grand National
Days in April, June, September and November

Shipping areas
Thirty something

31
Days in January, March, May, July, August, October and December

32
Ells in a bolt (cloth)
Teeth in the human mouth
Cards for a game of Euchre
Phosphorus 32

Water freezes (°F)
Cards each in Bezique
Points on the compass

33
Gertie Lee (Housy Housy)
Amstel beer (trente trois)

(and a third) RPM of an LP

34
Terror-Byte

35
Gallons (UK) in a barrel of oil
Cubic feet in a 'displacement' ton

36
Inches in a yard	Gallons in a barrel (brewing)
Thirty six line bible	Bushels in a chaldron
Pots for a maximum break in snooker

37
Shakespeare's plays
Body heat in Celsius (98.6 in Fahrenheit)
Numbers on a roulette wheel (incl. zero)

38
Smith and Wesson

39
The Thirty Nine Steps (John Buchan)
Books of the Old Testament (AV)
Articles of the Church of England ('forty stripes save one')
Steps to the royal box at the old Wembley Stadium

40
Days and nights in Lent	Poles in a furlong
Square poles in a rood	Double top or tops (darts)
Forty winks	Thieves with Ali Baba
Elijah fed by ravens for 40 days	Life begins at forty
Biblical period for embalming	Days for Nineveh to repent
Jesus' fast (days)	Ruby Wedding
40-40 = deuce in tennis
Spaces on a Monopoly board
Minutes to girdle the Earth (Puck)
UB 40 (Benefit form and Pop Group)
Forty Years On (Harrow Football Song)
Moses on the Mount for days and nights

Days before Noah opened the windows of the Ark
Weather predicted by St Swithin (days after July 15th)
Cents more for the next three minutes (*Sylvia's Mother*)

41
Bagooska the Terribly Tired Tapir

42
Gallons (US) in a barrel of oil   Forty two line bible
Lowest possible sum of scores at snooker (without fouls)
The Ultimate Answer (*Hitchhikers Guide to the Galaxy*) - actually 6x9!

43
Life Change Units in the Holmes and Rahe Scale

45
Colt .45   RPM of a 7" vinyl single
The second Jacobite Rebellion   45" = 1 ell (cloth)
P45 – pay and tax from previous employment

46
Chromosomes in a human cell (23x2)

47
Wigs of Grigory Griggs

48
Cards in pinochle

49
'49er' - Miner
PC 49 (in the Eagle and Brian Reece)

50

Bull's-eye (darts) Golden Jubilee/Wedding
Hawaii (50th state) Argonauts
Fifty-fifty (evens) Hawaii Five-oh
Meters length for an Olympic pool
Stars in the US flag (number of states now)
Bonus points for using all 7 tiles at scrabble

51

Will there ever be another State?

52

Cards in a pack Weeks in a year

54

Gallons in a hogshead
Cards in a pack (deck) including jokers

55

Fifty Five Days in Peking Emerald anniversary

56

Cards in the Tarot Minor Arcana

57

Heinz varieties
Fifty Seven Chevrolet (Dolly Parton)

58

Faces on a 'brilliant' cut diamond

59

Minutes from London to Brighton by train

60

Minutes in an hour	Minutes in a degree
Seconds in a minute (time)	Seconds in a minute (angle)
Metre dash	Minims in a fluid drachm
Diamond Jubilee/wedding isotope	Cobalt 60 - radioactive
	Angle of reflection for water
Nautical miles in a degree	P60 – tax record
Feet in a ten pin bowling lane	
Treble 20 - highest single dart score	
Sexagesima Sunday (60 days to Lent)	

61
Bonus squares in scrabble

62
Tickety boo (bingo)

64
Squares on a chessboard and draughtsboard

65
Number of Companions of Honour at any one time

66

Books of the Bible (AV)	*Route 66* (TV series)
(Get Your Kicks on) Route 66 (Bobby Troup)	

69

Vat	Soixante-neuf

70
Three score years and ten   Platinum anniversary
Mph UK national speed limit
Septuagesima (3rd Sunday before Lent)

72
Holes in a pro-golf tournament   Points to the inch in printing

73
Spaces on a Trivial Pursuit board

74
(or 64) guns on a third rate ship of the line

75
Diamond Wedding/Jubilee   cl of wine in a bottle

76
Trombones

77
Sunset Strip

78
RPM for an old single   Cards in a Tarot pack

79
79, Park Lane (Harold Robbins)

80
Minutes in a Rugby match   Chains in a mile
Around the World in 80 days

ZX80 Clive Sinclair's early personal computer

81
ZX81 - successor to the 80

84
84, Charing Cross Road (book and film)

86
Agent Maxwell Smart

88
Keys on a piano

90

Years without slumbering   Minutes in a soccer match
Guns on a second rater (up to 98)
Degrees in a right angle (quadrant)
Strontium 90 - radioactive isotope

95
Theses of Martin Luther

99

Ice cream and chocolate flake   Years for a normal lease
Bradman's test average (99.4)

100

A ton   Years in a century
Boiling point of water (Celsius)   Links in a chain
Cents in a dollar etc.....   Pence in a pound
Square metres in an 'are'   Hectares in a square km

Tiles in Scrabble

Game at bridge

'Hundred' = 100 hides

Senators in Congress

The Chiltern Hundreds

Napoleon's 100 days

Great Sporting Moments (BBC)

Days asleep for *The Sleeping Beauty*

Thebes – The Hundred Gated City

Cubic feet in a 'register' ton (shipping)

100+ guns on a first rate ship of the line

The Hundred Years War (England and France)

Eyes of Argus (and hence in the Peacock's tail)

101

101 Dalmatians (film)   *Room 101* (BBC programme)

102

102 Dalmatians (sequel to 101)

103

Flight number of Pan Am - Lockerbie disaster

105

North Tower (Tale of Two Cities - Dickens)

110

Voltage in some countries

112

Pounds in a hundredweight   Cell phone emergency

114

Chapters of the Koran

120
Fathoms in a cable (US) Feet in a bolt (cloth)

125
Intercity train

128
Cubic feet in a cord (timber)

131
Iodine 131 Radioactive isotope

136
Tiles in a basic Mah Jong set

137
Caesium 137 Radioactive isotope

143 "I love you" (number of letters)

144
Square inches in a square foot Tiles in a full Mah Jong set

147
Maximum break (and score) in snooker

149
BA flight number- landed in Kuwait at the worst possible time

150
Psalms in the Bible (AV)

153
International directory enquiries

168
Pips in a domino set

180
Highest three dart score (3x treble 20)
Degrees longitude for the dateline (approx.)
Degrees in a semi-circle

187
Years asleep for the Seven Sleepers

192
Directory enquiries

198
Radio 4 (long wave)

200
Pounds (dollars) for passing Go   Laps in the Indy 500

206
Bones in the human body

208
Radio Luxembourg frequency

212
Water boils (°F)

220
Yards in a furlong

220/240
Volts AC voltage in some countries

225
Improved Intercity train   Squares on a scrabble board

238
Uranium 238 - radioactive isotope

240
Old pennies in a pound   Yards in a cable (US)

245
Pounds weight for Big Bad John

247
Radio 1 on MW (not now)

273
Absolute zero (minus ºC)

286
Computer chip

300
Maximum in ten pin bowling   Spartans at Thermopylae

303 Lee Enfield

322 On skull and crossbones of undergrad society at Yale

357 .357 Magnum

360 Degrees in a circle

361 Points on a Go board

386 Faster computer chip

400
The 400 of ancient Athens   Forbes 400 (rich list)
Atari 400 computer
'The four hundred' - of New York Society in late 19th C

405 Lines on the original B&W TV

409 Bridges in Venice (reputedly)

418 Abrahadabra

420 4/20 cannabis culture

435 Representatives in Congress

444 American hostages held in Iran (days)

451 Spontaneous ignition of paper in Fahrenheit

461 *Ocean Boulevard* (Eric Clapton album)

470 Olympic class dinghy

480 Grains in an ounce

486 Even faster computer chip

500
Monkey = 500 pounds	Indianapolis 500
Sheets in a ream	500 Rum (rummy variation)
Daytona 500 etc…	Fiat 500

500 Miles by Bobby Bare, Peter Paul and Mary et al

I'm gonna be (500 Miles) The proclaimers
Five Hundred Points of Good Husbandrie (Thomas Tusser)

501 Game at championship darts   Levi Strauss jeans

505 U-505   Levi jeans

525 Lines on a (US) NTSC TV

535 Abb. for Tiananmen Square protests of 1989

555 State Express cigarettes

562 Number of Native American Nations

586 Chip that became the 'Pentium'

600
Into the valley of death rode the 600 (Tennyson)
Roman soldiers in a cohort

608 Feet in a cable (UK)

625
Lines on a later B&W TV
Square links in a square rod

617 Squadron - "The Dambusters"

640 Acres in a square mile

650 MPs in the House of Commons (2017)

666 The mark of the beast

693 Radio 5 live on MW

700 *The 700 Club* on CBN (Christian Broadcasting Network)

707 Noisy old Boeing (4 engines)

714
Home runs by Babe Ruth   Joe Friday's badge number
Flight 714 to Sidney (Tintin novel)

717 Almost missing Boeing

727 Another Boeing (3 engines)

720 Feet in a cable (US)

737 Very common Boeing (2 engines)

747 Jumbo Jet Boeing (4 engines)

750 ml of wine in a bottle

757 Thin boring Boeing (2 engines)
'The 757' Hampton Roads in Virginia

767 Fat boring Boeing (2 engines)

770 *File 770* Sci-fi Fanzine

777
Much newer Boeing (2 engines)   Perfect number
Triskelion on the AWB (Afrikaner Resistance Movement)

787 Boeing Dreamliner (2 engines)

800 Atari 800 computer

804 Greater Richmond in Virginia

888 Ceefax subtitles (no longer available)

893 Considered unlucky in Japan

900 Skateboard spin of two and a half times

905 Greater Toronto

909 Radio 5 live on MW

911
Popular Porsche
Emergency telephone number (USA)

913 Obi-Wan Kenobi emergency transmission

925 Sterling silver

958 Britannia silver (alloyed with copper)

962 Age of Jared - second oldest after Methuselah

969 Age of Methuselah

999 Emergency services in UK   London based punk band

1000
k	Days of JFK's presidency
Metres in a kilometre	litres in a cubic metre
Grams in a kilogramme	kilograms in a tonne

A grand
Years in a millennium
One in a thousand
Anne of a 1000 Days (Boleyn)
Length of the Mille Miglia in km

Guineas (Classic horserace)
Watts in a kilowatt
A picture is worth in words

1001 Arabian nights   Carpet cleaner

1064 Real value of k (commonly used for 1000)

1066 *1066 and All That* (book by Sellar and Yeatman)

1210 Square yards in a rood

1212 Scotland Yard (the old Whitehall exchange number)

1343 Height of Ben Nevis (m)

1400
'Fourteen hundred' - strangers present in the Stock Exchange

1440 Minutes in a day

1471 Who phoned you last?

1500 Metric mile (m)
Pounds to start at Monopoly
The Light Programme (metres wavelength)

1600 Pennsylvania Avenue: The White House

1664 Krönenberg Beer

1728 Cubic inches in a cubic foot

1760 Yards in a mile

1812 Overture (Tchaikovsky)

1852 Metres in a nautical mile

1922 Committee (Conservative politics)

1976 Yards in an Ascot mile

1984 Novel by George Orwell

2000 Guineas (Classic horserace)
Pounds in a tonne (US and aviation)

2001 *2001 A Space Odyssey* - film by Stanley Kubrick

2040 *Space Precinct* (TV series)

2182 KHz - the emergency frequency

2240 Pounds in a ton

2455 *Cell 2455 on Death Row* (by Caryl Chessman)

3000 Feet for a 'Munroe' (hill-climbing)
Meters in a steeplechase

3600 Seconds in an hour

4000 Fed with seven loaves and a few fishes

4040 Four star petrol

4077 *M*A*S*H*

4711 Eau de Cologne

4840 Square yards in an acre

5000 Game in Canasta
Fed with 5 loaves and 2 fishes

5280 Feet in a mile

6080
Feet in a nautical mile (now 6,076.1 by international standard)

7070 Lead free petrol

7927 Equatorial diameter of the Earth in Miles

8086 First computer chip

8848 Height of Everest in metres

9000 HAL (the errant computer in *2001 A Space Odyssey*)

10,000 Myriad
Square metres in a hectare
Men with the Grand Old Duke of York

20,000 Leagues under the sea

24,902 Distance round the equator in miles

29,028 Height of Everest in feet

40,000 Men with the King of France
Feathers on a thrush

64,000 Dollar question

65000 *Pennsylvania 65000* (Glenn Miller)

70,000
Heads, faces, mouths, tongues and languages in seventh heaven

86,400 Seconds in a day

90210 *Beverley Hills* (the TV series)

99,966 Genes in the human body

100,000 A lakh (India)

1 million Mega (10^6)   Microns in a metre
Years BC (Raquel Welch's film)

6 Million *Dollar Man* (starring Lee Majors)

8 Million stories in *The Naked City* (Jules Dassin)

10 million A crore (India)

1,000 million Giga (10^9)
A billion (US and general usage)
also known as a milliard

1 million million
A billion (UK - not now in general use) Tera (10^{12})

Googol 10 to the power 100

Googolplex 10 the power of a googol

ORDINALS

1st

First Among equals

First Class

First (class honours degree)

First born

First degree burn

First foot(ing)

(At) first hand

First name

First mate

First person (grammatical)

First refusal

First water (diamond)

First Fleet(er) - to Australia

First light

First USA state – Delaware

First rater (100+ gun warship)

First Violin - leader of an orchestra

The First Jacobite rebellion (1815)

First Gentleman of Europe (George IV)

First amendment (concerning freedoms)

First night(er)

First thing

First aid

First day cover

First degree murder

First degree

First lady

First rate

First reading

First officer

First strike

First come first served

The First Letter (Kipling)

First-fruit

First offence (or offender)

First past the post

2nd

Second wind

Second class

Second cousin

Second hand

Second coming

Second degree burn

Second empire
Second rate
Second childhood
Second nature
Second string
Second guessing
Second fiddle
Second generation
Second growth
Second sight
Second thoughts
Second rater (90 gun warship)
The Second Jacobite rebellion - 1845
Second Amendment (right to bear arms)
Second chamber (e.g. The Lords or the Senate)
The Adventure of the Second Stain (Sherlock Holmes)

3rd
Third degree
Third world
The Third Man
Third dimension
Third rate
Third party
Third age
Third class
Third eyelid
Third man (cricket)
Third Order (Monks)
Third person (grammatical)
Close Encounters of the Third Kind (film)
Third rater (64 or 74 gun warship)

4th
Fourth dimension (time)   Fourth Estate (the Press)
Fourth of July - Independence Day in the USA

5th
Fifth column
Fifth wheel (the spare)
Fifth generation computer
Fifth Estate (The BBC)

Fifth amendment (the right not to incriminate yourself)

6th
Sixth formSixth sense
Inn of the Sixth Happiness (book and film)

7th
Seventh heavenThe Sabbath
Seventh son of a seventh sonSeventh Day Adventist
The Seventh Scroll (Wilbur Smith)
Seventh Kingdom of the Antichrist (Revelations)

8th
Eighth Army - the Desert Rats

9th
Planet Beyond Neptune but not Pluto which has been downgraded. If Pluto is included as the ninth, then it becomes the 'tenth planet'

11th
At the eleventh hourEleventh chord (jazz)
Eleventh Commandment - 'thou shall not be found out'
Eleventh hour of the eleventh day (Armistice WWI)

12th
Twelfth NightTwelfth Day
Glorious Twelfth (August)Twelfth man (cricket)

13th
Ides of Jan, Feb, Apr, Jun, Aug, Sep, Nov, Dec
Friday the 13th

15th	Ides of March, May, July and October
17th	Parallel between North and South Vietnam
18th	Birthday - entitled to vote in the UK
19th	'Province of Iraq' (Kuwait)
20th	20th Century Fox
21st	Birthday - key of the door
23rd	*Psalm* ('The Lord's my Shepherd')
38th	Parallel - divides Korea
42nd	Street in Manhattan and musical based on it
49th	Parallel between USA and Canada
50th	State of the USA – Hawaii
97th	Bomb disposal unit in 'Danger-UXB'
256th	Capt Yossarian's US Army Corps

GROUPS

Historical One eyed characters

- The Cyclops (Greek mythology)
- Davy Gam (knight at Agincourt)
- Graeae (Greek myth)
- Hannibal
- Horatius
- Marshland Shales (19th C horse)
- Horatio Nelson
- John Milton
- Samuel Johnson (dictionary)
- Odin or Woden/Wotan
- Polyphemus (a Cyclop)
- Philip of Macedon
- Wackford Squeers (Nicholas Nickleby)

Persons who have only one eye in more modern times

- Gordon Banks (goalkeeper)
- Joe Davis (snooker)
- Moshe Dayan (Israeli general)
- John Ford (film director)
- Rex Harrison (actor)
- Colin Milburn (cricketer)
- Herbert Morrison (politician)
- James Thurber (writer)
- Sammy Davis Jr
- Peter Falk (actor)
- Leo Fender (guitars)
- Theodore Roosevelt (president)
- Sandy Duncan (The Hogan Family)
- Bushwick Bill (rapper)

- Kirby Puckett (baseball)

Some one-armed persons

- Nelson (admiral)
- Captain Hook (*Peter Pan*)
- Stonewall Jackson (general)
- Cervantes (writer)

Some one-legged persons and characters

- Cole porter (composer)
- Sarah Bernhardt (actress)
- Al Capp (cartoonist)
- Peter Stuyvesant (New Amsterdam)
- Captain Ahab (Melville)
- Berchta (German Myth)
- Long John Silver (*Treasure Island*)
- Silas Wegg (Dickens' *Our Mutual Friend*)

The Wetherfield One

- Dierdre Rachid (*Coronation Street*)

Two Sicilies (1061-1860)

- Sicily
- Naples

Two Gentlemen of Verona

- Valentine
- Proteus

The Two Sacraments (see also 5 and 7)

- Babtism
- The Lord's Supper

The Dynamic Duo

- Bruce Wayne   (Batman)
- Dick Grayson   (Robin)

The Two Cultures

- Humanities
- Science

The Tamworth Two (pigs)

- Butch
- Sundance

The Two Eyes of Greece

- Athens
- Sparta

The Twins

- Castor
- Pollux

The Pillars of Hercules

- Rock of Gibraltar
- Jubal Musa

The Merry Wives of Windsor

- Mistress Ford
- Mistress Page

The Three Wise Men (or Kings - The Magi)

- Melchior   Gold
- Gaspar   Frankincense
- Balthazar   Myrrh

Three Virtues

- Faith
- Hope
- Charity

Three Graces

- Aglaia
- Euphrosyne
- Thalia

Three Muses

- Aoide   Song
- Melete   Meditation
- Mneme   Remembrance

The Sirens

- Leucosia
- Ligea
- Parthenope

The Three Graces of Liverpool

- Royal Liver Building
- Cunard Building
- Port of Liverpool Building

The Christian Trinity

- God the Father
- God the Son
- God the Holy Ghost (Holy Spirit)

Funeral Tolls

- 3 for a child
- 6 for a woman
- 9 for a man

The Three Orders

- 1st   Friars
- 2nd   Nuns
- 3rd   Order of Monks striving for perfection

The Hindu Trinity (Trimurti)

- Brahma the creator (consort – Sarasvati/Kali)
- Vishnu the sustainer (consort – Lakshmi)
- Shiva the destroyer (consort – Parvati)

Three Teachings (Harmonious as One)

Confucianism, Taoism and Buddhism

Three Estates

- Lords Spiritual
- Lords Temporal
- Commons

Three Fates

- Clotho
- Lachesis
- Atropos

Three Furies

- Tisiphone
- Alecto
- Megaera

Three Harpies

- Aello
- Celeno
- Ocypete

The Three Norns

- Past   Urda
- Present   Verdandi
- Future   Skuld

The Gorgons

- Euryale
- Medusa
- Sthena

The Cyclops

- Arges
- Brontes
- Steropes

Three-day Eventing

- Dressage
- Cross Country (Endurance)
- Show Jumping

The Triple Crown

- Kentucky Derby
- Preakness Stakes
- Belmont Stakes

The Triathlon (Olympic or International Distance)

- Swimming 1.5 km
- Cycling 40 km
- Running 10 km

The Three 'R's

- Reading
- (W)riting
- (A)rithmetic

Freud's Stages of Childhood Development

- Oral
- Anal-sadistic
- Phallic

Freud's Parts of the Psyche

- Ego
- Id
- Superego

Three 'F's of Ireland

- Fair rent
- Free sale
- Fixity of tenure

The Three Tongues of the Crucifixion

- Hebrew
- Greek
- Latin

Three Inalienable Rights

- Life
- Liberty
- The Pursuit of Happiness

Three Counties (Choir and Show)

- Hereford(shire)
- Worcester(shire)
- Gloucester(shire)

The Three Musketeers

- Athos
- Porthos
- Aramis

The Three Tenors

- Luciano Pavarotti
- Placido Domingo
- Jose Carreras

The Primary Colours

- Red
- Blue
- Green

The Secondary (subtractive) Colours

- Cyan
- Magenta
- Yellow

Three Kingdoms of Nature

- Animal
- Vegetable
- Mineral

Three Men in a Tub

- Butcher
- Baker
- Candlestick maker

Three Men in a Boat

- Harris
- George
- The Narrator ("I")

Close Encounters

- Sighting
- Physical evidence
- Contact

The Trivium (roads to learning)

- Grammar
- Rhetoric
- Logic

The Three Stooges

- Moe Howard
- Larry Fine
- Curly Howard (also Shemp Howard)

The Three Little Pigs

- Fifer Pig   straw house
- Fiddler Pig   wooden house
- Practical Pig   brick house

The Four Last Things

- Death
- Judgement
- Heaven
- Hell

The Four Gospels

- Matthew
- Mark
- Luke
- John

The Four Heads on Mount Rushmoor

- Washington
- Jefferson
- Lincoln
- Roosevelt

The Four Causes of Aristotle

- Material cause
- Substantial cause
- Efficient cause
- Final cause

Four 'Withouts'

- Manners
- Wit
- Money
- Credit

The Four Horsemen of the Apocalypse

- White Horse   Power of God   (Bow and Crown)
- Red Horse   Bloodshed and War   (Sword)
- Black Horse   Famine   (Scales)
- Pale Horse   Death and Disease

The Four Masters (Annals of the Kingdom of Ireland)

- Michael O'Clery
- Conaire O'Clery
- Cucoigriche O'Clery
- Fearfeasa O'Mulconry

The Grand Slam Tennis Tournaments

- Australian
- French
- US
- Wimbledon

Gang of Four (China)

- Madame Mao
- Yao Wenyuan
- Zhang Chunqiao
- Wang Hong-wen

Gang of Four (UK)

- Roy Jenkins
- Shirley Williams
- William Rodgers
- David Owen

Four Kings of a French Pack

- Charlemagne
- David
- Alexander the Great
- Caesar

The Four Freedoms of FDR

- Freedom of Speech
- Freedom of Expression
- Freedom in Worship
- Freedom from Fear and Want

The Four Marys (companions of Mary Stewart)

- Mary Beaton (Bethune)
- Mary Livingstone (Leuson)
- Mary Fleming (Flymyng)
- Mary Seaton (Seyton)

The Four Thomas's (Chancellors of Henry VIII)

- Thomas Wolsey — died
- Thomas More — executed
- Thomas Cromwell — executed
- Thomas Cranmer — survived but executed by Mary Tudor

The Four Balls in Croquet

- Black and Blue play Red and Yellow

The Four Quarter Days

- Lady Day — March 25
- Midsummer — June 24
- Michaelmas — September 29
- Christmas — December 25

Four Quartets of TS Eliott

- Burnt Norton
- East Coker
- The Dry Salvages
- Little Gidding

Four US Generals who later became Presidents

- Ulysses S Grant
- Andrew Jackson
- George Washington
- Dwight D Eisenhower

The Four Rivers of Creation (Genesis)

River	Land
Pishon	Havilah
Gihon	Cush
Hiddekel	Assyria
Euphrates	

The Fantastic Four (DC Comics)

- Mr Fantastic
- Invisible Girl
- The Human Torch
- The Thing

The Four Sons of Aymon

- Renauld (Rinaldo)
- Alard
- Guichard
- Richard

The Big Four Sporting Events

The Summer Olympics
The Winter Olympics
The World Athletic Championships
The World Cup (Soccer)

Germany, France, Italy and Japan have held all of them and will be joined by South Korea and Russia in 2018.

The Five Nations (American Indian Confederacy - collectively the Iroquois)

- Cayugas
- Mohawks
- Oneidas
- Onondagas
- Senecas

The Five Marx Brothers

- Groucho   (Julius)
- Harpo   (Arthur or Adolf)
- Zeppo   (Herbert)
- Chico   (Leonard)
- Gummo   (Milton)

The Five Classics of Horseracing

- Derby
- Oaks
- St Leger
- One Thousand Guineas
- Two thousand Guineas

The Olympic Pentathlon (for women - in Olympics till replaced by Heptathlon in 1981)

- 800 metres
- 100 metre hurdles
- high jump
- long jump
- shot

Events of the Modern Pentathlon

- Riding
- Fencing
- Shooting
- Swimming
- Running

The Five Classic Orders of Architecture

- Composite
- Corinthian
- Doric
- Ionic
- Tuscan

The Five K's of Sikhism

- Kanga - comb
- Kara – steel wrist band
- Kesh – uncut hair
- Kirpan - sword
- Kaccha - trousers

Five Tastes

- Sweet
- Sour
- Salt
- Bitter
- Savoury (umami)

The Famous Five (Enid Blyton)

- Julian
- George (Georgina)
- Ann
- Dick
- Timmy (the dog)

The Pentateuch

- Genesis
- Exodus
- Leviticus
- Numbers
- Deuteronomy

The Five Agents of Yin and Yang

- Metal
- Wood
- Water
- Fire
- Earth

The D Day Beaches

- UTAH   American 7th Corps
- OMAHA   American 5th Corps
- GOLD   British 30th Corps
- JUNO   British 1st Corps and Canadians
- SWORD   British 1st Corps

Five Members of the Long Parliament (1642)

- Pym
- Hampden
- Haselrig
- Holles
- Strode

The Five Senses

- Touch   (somatosensation or tactition)
- Taste   (gestation)
- Smell   (olfaction)
- Sight   (vision)
- Hearing   (audition)

The sixth sense is generally accepted to be extra-sensory perception but why are the senses of: being able to feel heat (thermoception); pain (nociception) or time (chronoception) not normally listed?

The Five Permanent Members of the UN Security Council

- China
- France
- Russia
- United Kingdom
- United States

The Five Wits

- Common Sense
- Imagination
- Fantasy
- Estimation
- Memory

The 'Five Towns' of Arnold Bennett

- Tunstall
- Burslem
- Fenton
- Hanley
- Longton

(Which makes six if you include Stoke)

The Cinque Ports (Whilst cinque means five there are actually seven of them depending on when you choose to count them)

- Hastings
- Sandwich
- Dover
- New Romney
- Hythe

Plus Winchelsea and Rye

The Five old English Counties which disappeared under the 1972 Act

- Huntingdonshire
- Cumberland
- Middlesex
- Westmoreland
- Rutland (reinstated 1997)

The Five Boroughs of the Danish Confederation

- Derby
- Leicester
- Lincoln
- Nottingham
- Stamford

The Five Articles of Perth 1618

- restore Bishops
- kneel for communion
- private communion if necessary
- observation of Feast Days
- confirmation in church

The Great Lakes

- Superior
- Michigan
- Huron
- Erie
- Ontario

The Great Lakes of East Africa

- Rudolph
- Albert
- Victoria
- Tanganyika
- Nyasa

Five Sacraments

- Confirmation
- Penance
- Orders
- Matrimony
- Extreme unction

The Five Points of Calvinism

- Transcendence of God
- Depravity of natural man
- Predestination and election for salvation
- Authority of the Scriptures
- Enforcement of the Church's discipline

Kipling's Six Nations

- The Five Nations of the Iroquois plus
- The Tuscaroras

The Six Wives of Henry VIII

- Catherine of Aragon (divorced)
- Anne Boleyn (beheaded)
- Jane Seymour (died)
- Anne of Cleves (divorced)
- Catherine Howard (beheaded)
- Catherine Parr (survived)

Six Nations Championship, Rugby Union

- England
- Scotland
- Ireland (both Northern and Republic combined)
- France
- Wales
- Italy

The Six (Original Common Market)

- Belgium
- France
- Germany
- Italy
- Netherlands
- Luxembourg

Les Six (French Composers)

- Honnegar
- Milhaud
- Poulenc
- Durey
- Auric
- Tailleferre

Six Honest Serving Men (Kipling)

- What?
- Why?
- When?
- How?
- Where?
- Who?

The Six Counties (of Northern Ireland)

- Fermanagh
- (London)Derry
- Antrim
- Down
- Armagh
- Tyrone

Six Archangels of the Bible

- Gabriel
- Raphael
- Uriel
- Chamuel
- Jophiel
- Zadkiel

The Six Acts

- Restriction of public meetings
- Speedier trials
- Forbids drilling by private persons
- Seizure of blasphemous and seditious literature
- Seizure of arms
- Stamp duty on newspapers

The Six Articles ('The Bloody Bill')

- Transubstantiation
- Sufficiency of Communion
- Clerical celibacy
- Obligation of monastic vows
- Propriety of private masses
- Necessity of auricular confession

The Six Articles of English Ritualism

- Altar lights
- Eucharistic vestments
- Mixed chalice
- Incense
- Unleavened bread
- Eastward position

The Seven Virtues

- Faith
- Hope   (The three virtues)
- Charity
- Justice
- Fortitude   (Plato's cardinal virtues)
- Prudence
- Temperance

The Seven Churches of Asia

- Ephesus
- Smyrna
- Pergamos
- Thyatria
- Sardis
- Philadelphia
- Laodicea

The Seven Sleepers

- Constantius
- Dionysius
- Joannes
- Maximianus
- Malchus
- Martinianus
- Serapion

The 'Seven Sisters' of the Oil Industry

- BP
- Gulf
- Mobil
- Shell
- Texaco
- Standard Oil of California
- Standard Oil of New Jersey

The Magnificent Seven

- Yul Brynner   Chris   (survived)
- Charles Bronson OReilly   (killed)
- Horst Buckholz   Chico   (killed)
- Steve McQueen   Vin   (survived)
- Robert Vaughn   Lee   (killed)
- James Coburn   Britt   (killed)
- Brad Dexter   Harry   (survived and stayed)

The Seven Samurai

- Kambei
- Kikuchingo
- Gorobei
- Kyunzo
- Heihachi
- Shichiroji
- Katsushiro

The Seven Dwarfs

- Sleepy
- Doc (is in charge)
- Bashful
- Happy
- Dopy
- Grumpy
- Sneezy

Seven principal dancing movements in ballet

- tourner (to turn)
- glisser (to glide)
- sauter (to jump)
- élancer (to dart)
- plier (to bend)
- étendre (to stretch)
- rélever (to rise)

The Seven Sages

- Bias (of Priene)
- Chilo (of Sparta)
- Cleobulus (of Lindus)
- Periander (of Corinth)
- Pittacus (of Mitylene)
- Solon (of Athens)
- Thales (of Miletus)

The Seven Deacons

- Stephen
- Philip
- Prochorus
- Nicanor
- Timon
- Parmenas
- Nicholas

The Seven Deadly Sins

- Pride   (Vainglory)
- Avarice   (Covetousness)
- Lust
- Envy
- Gluttony   (Greed)
- Anger   (Wrath)
- Sloth

Seven Gifts of the Spirit

- Wisdom
- Understanding
- Counsel
- Fortitude
- Knowledge
- Righteousness
- Fear of the Lord

The Seven Days of Creation

- Light, night and day
- Firmament/heaven
- Earth, seas and plants
- Sun and moon
- Fish and fowl
- Beasts and man
- Rest

The Seven Wonders of the (ancient) World

- The Pyramids of Egypt   Egypt
- The Statue of Zeus (Jupiter) at Olympia   Greece
- The Tomb Of Mausolus at Helikarnasus   Bodrum
- The Colossus of Rhodes   Greek Island
- The Hanging Gardens of Babylon   Iraq
- The Pharos of Alexandria   Egypt
- The Temple of Diana (Artemis) at Ephesus   Selkuk

The Seven Wonders of the Middle Ages

- The Coliseum of Rome
- The Catacombs of Alexandria
- The Great Wall of China
- Stonehenge
- The Leaning Tower of Pisa
- The Porcelain Tower of Nanking
- The Mosque of St Sophia in Constantinople

The Seven Cities claiming to be the Birthplace of Homer

- Argos
- Athens
- Chios
- Colphon
- Rhodes
- Salamis
- Smyrna

The Seven Seas

- North Pacific
- South Pacific
- North Atlantic
- South Atlantic
- Arctic Ocean
- Indian Ocean
- Antarctic or Southern Ocean

(Antarctica had not been discovered)

The Liberal Arts (of Mediaeval Universities)

- Grammar
- Rhetoric   (The Trivium)
- <u>Logic</u>
- Arithmetic
- Geometry   (The Quadrivium)
- Astronomy
- Music

The Seven Senses of the Soul

- Fire gives animation
- Earth gives feeling
- Water gives speech
- Air gives taste
- Mist gives sight
- Flowers give hearing
- The South Wind gives smell

The Seven Hills of Rome

- Palatine
- Capitoline
- Quirinal or Colline
- Caelian
- Aventine
- Esquiline
- Viminal

The Seven Japanese Gods of Luck

- Benten   Goddess of Love
- Bishamon   War
- Daikoku   Wealth
- Ebisu   Self-effacement
- Fukurokuji   Longevity
- Jojorm   Longevity
- Hotei   Joviality

The Seven Joys of Mary

- Annunciation
- Visitation
- Nativity
- Epiphany
- Finding in the Temple
- Resurrection
- Ascension

The Seven Sorrows of Mary

- Simeon's prophecy
- Flight into Egypt
- Loss of the Holy Child
- Meeting the Lord on the way to Calvary
- The crucifixion
- Taking down from the cross
- Entombment

Seven Works of Corporal Mercy

- Feed the hungry
- Clothe the naked
- Tend the sick
- Bury the dead
- Give drink to the thirsty
- Take in strangers
- Minister to prisoners

The Seven Spiritual Works of Mary

- convert the sinner
- instruct the ignorant
- counsel those who doubt
- comfort those in sorrow
- bear wrongs patiently
- forgive injuries
- pray for the living and the dead

The Seven Champions of Christendom

- St George   (England)
- St Andrew   (Scotland)
- St David   (Wales)
- St Patrick   (Ireland)
- St Denys   (France)
- St James   (Spain)
- St Anthony   (Italy)

The Seven Bishops

- Archbishop Sancroft of Canterbury
- Bishop Lloyd of St Asaph
- Bishop Turner of Ely
- Bishop Ken of Bath and Wells
- Bishop White of Peterborough
- Bishop Lake of Chichester
- Bishop Trelawney of Bristol

Seven Plagues of Seven Angels (Revelations, Ch 16)

- Sores
- Sea turned to blood
- Rivers turned to blood
- Men scorched with fire
- Darkness
- Euphrates dries up
- Lightning, thunder and earthquake

Seven Fathers of the Church

- St Athenasius
- St Gregory of Nazianzen
- St John Chrysostom
- St John of Damascus
- St Basil of Caesarea
- St Gregory of Nyssa
- St Cyril of Alexandria

The Seven Sacraments (see also 2 and 5 sacrements)

- Baptism
- Confirmation
- Eucharist
- Penance
- Orders
- Matrimony
- Extreme unction

Seven against Thebes

- Adrastus (Leader and sole survivor)
- Polynices
- Amphiaraus
- Capaneus
- Hippomedon
- Tydeus
- Parthenopaeus

The Seven Ages of Man (Shakespeare - *As You Like It*)

- Infant
- Schoolboy
- Lover
- Soldier
- Justice
- Pantaloon
- Second childhood

Seven Soldiers of Victory (DC Comics)

- Crimson Avenger
- Green Arrow
- Speedy
- Shining Knight
- Star-spangled Kid
- Stripsey
- Vigilante

The Secret Seven (Enid Blyton)

- Peter and Janet
- Jack and Barbara
- George and Pam
- Colin

The G7 Nations

- Japan
- USA
- Germany
- France
- Italy
- Canada
- UK

The G8 Nations

- G7 plus
- Russian Federation

Father Christmas's Reindeer

- Dancer and Prancer
- Donner and Blitzen
- Vixen and Comet
- Cupid and Dasher

(Rudolph is not one of the officially listed first team)

The Eight (new) Welsh counties

- Gwynned
- Dyfed
- Gwent
- Clwyd
- Powys
- West Glamorgan
- Mid Glamorgan
- South Glamorgan

The Noble Eightfold Path of Buddhism

- Right view
- Right thought
- Right speech
- Right action
- Right livelihood
- Right effort
- Right mindfulness
- Right concentration

The Ivy League

- Brown
- Columbia
- Cornel
- Dartmouth
- Harvard
- Princeton
- Pennsylvania
- Yale

The Nine Orders of Angels

- Seraphim (winged head of child)
- Cherubim (winged child)   [First Circle]
- Thrones
- Dominations
- Virtues   [Second Circle]
- Powers
- Principalities
- Archangels   [Third Circle]
- Angel

The Nine Orders of Knighthood

- The Garter (KG)
- The Thistle (KT)
- St Patrick (KP)
- The Bath (KB)
- The Star of India
- St Michael and St George (KMG)
- The Indian Empire
- Royal Victorian Empire
- The British Empire

The Nine Worthies

- Joshua
- David   (three Jews)
- <u>Judas Maccabaeus</u>
- Hector
- Alexander the Great   (three Pagans)
- <u>Julius Caesar</u>
- King Arthur
- Charlemagne   (three Christians)
- Geoffrey (or Godfrey) of Bouillon

The Nine Muses (the Daughters of Zeus and Mnemosyne)

- Calliope   Epic Poetry
- Clio   History
- Erato   Love Poetry
- Euterpe   Lyric Poetry and Music
- Melpomene   Tragedy
- Polyhymnia   Singing, Mime and Sacred Dance
- Terpsichore   Dance and Choral Song
- Thalia (also a grace) Comedy and Pastoral Poetry
- Urania   Astronomy

Nine Roman Forts of the Saxon shore

- Brancaster (Norfolk)
- Burgh Castle (Norfolk)
- Walton (Suffolk)
- Bradwell on Sea (Essex)
- Reculver (Kent)
- Richborough (Kent)
- Lympne (Kent)
- Pevensey (Sussex)
- Portchester (Hampshire)

The Nine Worthies of London

- Sir William Walworth
- Sir Henry Pritchard
- Sir William Sevenoke
- Sir Thomas White
- Sir John Bomham
- Christopher Croker
- Sir Harold Hawkwood
- Sir Hugh Calverley
- Sir Henry Maleverer

The Nine Points of Heraldry

- Dexter chief point
- Chief point (upper section)
- Sinister chief point
- Honor point
- Fesse
- Nombril
- Dexter base point
- Base point (lower section)
- Sinister base point

The Hollywood Ten

- Alvah Bessie
- Herbert Biberman
- Lester Cole
- Powys
- Ring Lardner, Jr.
- John Howard Lawson
- Albert Maltz,
- Samuel Ornitz
- Dalton Trumbo
- Adrian Scott

The Ten Commandments (the Decalogue)

- Worship only Me
- Graven Images
- Idols
- Name in Vain
- Sabbath
- Parents
- Murder
- Adultery
- False Accusation
- Desire

The Decathlon

- 100 metres
- 400 metres
- 1500 metres
- high jump
- 110 metres hurdles
- long jump
- javelin
- pole vault
- shot
- discus

The Ten Plagues of Egypt (Exodus)

- Water to Blood
- Frogs
- Lice, sand flies or fleas
- Swarms of flies
- Cattle die
- Boils and sores
- Hail
- Locusts
- Darkness
- Death of Firstborn

The Ten Avatars of Vishnu

- Matsya — Fish
- Kurma — Tortoise
- Varaha — Boar
- Nrisinha — Man/lion
- Vamana — Dwarf
- Parasurama — Rama with the Axe

- Ramachandra — Rama
- Krishna — Man
- Buddha — Man
- Kalki — White, winged horse

The Ten Pythagorean Opposites (after Aristotle)

- Limited — Unlimited
- Odd — Even
- Unity — Plural
- Right — Left
- Male — Female
- At rest — In motion
- Straight — Curved
- Light — Darkness
- Good — Evil
- Square — Oblong

The Twelve Labours of Hercules

- Slay the Nemean Lion
- Kill the Lernean Hydra
- Catch the Arcadian Stag (Ceryneian Hind)
- Destroy the Erymanthean Boar
- Cleanse the Augean Stables
- Destroy the Cannibal Birds of Lake Stymphalis
- Capture the Cretan Bull
- Catch the Thracian Horses
- Obtain the Girdle of Hippolyte (Queen of the Amazons)
- Capture the Oxen of Geryon
- Possess the apples of Hesperides
- Bring Cerberus up from Hades

The Twelve Tribes of Israel (named after the twelve sons of Jacob)

- Reuben (Manasseh)
- Simeon
- Levi
- Judah
- Dan
- Naphtali
- Gad
- Asher
- Issachar
- Zebulin
- Joseph (of the coat)
- Benjamin (Ephraim)

The Twelve Disciples/ Apostles

- Simon — aka Peter
- Andrew — brother of Simon
- James — son of Zebedee
- John — brother of James
- Philip
- Bartholemew — aka Nathaniel
- Thomas — 'doubting'
- Matthew
- James — brother of John
- Lebbaeus — aka Thadaeus or Judas son of James
- Simon — The Canaanite or The Zealot
- Judas Iscariot — replaced by Matthias

Whilst Paul is the best known of the Apostles, he is not listed in the twelve. Matthew and John have gospels attributed to them along with Mark and Luke.

Twelve Good Rules (attributed to King Charles I)

- Urge no healths
- Profane no divine ordinances
- Touch no state matters
- Reveal no secrets
- Pick no quarrels
- Make no comparisons
- Maintain no ill opinions
- Keep no bad company
- Encourage no vice
- Make no long meals
- Repeat no grievances
- Lay no wagers

The Dirty Dozen

- John Cassavete (Victor Franko)
- Charles Bronson (Joseph Wladislaw)
- Jim Brown (Robert Jefferson)
- Telly Savalas (Archer Maggott)
- Donald Sutherland (Vernon Pinkley)
- Clint Walker (Samson Posey)
- Trini Lopez (Pedro Jiminez)
- Tom Busby (Milo Vladek)
- Benito Carruthers (Glenn Gilpin)
- Stuart Cooper (Roscoe Lever)
- Colin Maitland (Seth Sawyer)
- Al Mancini (Tassos Bravos)

The Thirteen (to 2018) Doctors Who

1. William Hartnell
2. Patrick Troughton
3. Jon Pertwee
4. Tom Baker
5. Peter Davison
6. Colin Baker
7. Sylvester McCoy
8. Paul McGann
9. Christopher Eccleston
10. David Tennant
11. Matt Smith
12. Peter Capaldi
13. Jodie Whittaker

(Dr Who has also been played by Peter Cushing (twice) and by Paul McGann (once) in films)

The Best Known Knights of the Round Table

Sir Kay	Sir Gareth
Sir Bors	Sir Lancelot
Sir Galahad	Sir Tarquin
Sir Modred*	Sir Bediver
Sir Gawain	Sir Tristran de Lyonnais
Sir Percival	Sir Ector
Sir Lionel	

*Arthur's Son

Thirteen Welsh counties before reorganisation

Monmouth   Glamorgan
Carmarthen   Pembroke
Cardigan   Brecknock (Brecon)
Radnor   Montgomery
Merioneth   Caernarvon
Anglesey   Denbigh
Flint

Thirteen Original States of the USA

The New England Colonies
New Hampshire Province, chartered as British colony in 1679
Massachusetts Bay Province, chartered as a British colony in 1692
Rhode Island Colony, chartered as a British colony in 1663
Connecticut Colony, chartered as a British colony in 1662

The Middle Colonies
New York Province, chartered as a British colony in 1686
New Jersey Province, chartered as a British colony in 1702
Pennsylvania Province, a proprietary colony established in 1681
Delaware Colony (before 1776, the Lower Counties on the Delaware River), a proprietary colony established in 1664

The Southern Colonies
Maryland Province, a proprietary colony established in 1632
Virginia Dominion and Colony, a British colony established in 1607
Carolina Province, a proprietary colony established 1663
North and South Carolina divided provinces, each chartered as British colonies in 1729
Georgia Province, a British colony established in 1732

The Fourteen Stations of the Cross

1. Condemned
2. Bears cross
3. Falls first time
4. Meets Mother
5. Simon helps
6. Veronica wipes face
7. Falls second time
8. Consolation
9. Falls third time
10. Unclothed
11. Nailed to cross
12. Dies
13. Taken down
14. Entombed

The Fourteen steps of Alcoholics Anonymous

The original wording has been foreshortened below:

Admit powerlessness
Believe in greater power
Decide to empower God
Make moral inventory
Admit nature of wrongs
Be ready to remove defects
Ask to remove shortcomings
List those harmed
Make amends
Take stock
Pray and meditate
Awaken, share and practice

Foyol's Fourteen Principles of Management

1. Division of Work
2. Authority and Responsibility
3. Discipline
4. Unity of Command
5. Unity of Direction
6. Subordination of Individual Interest
7. Remuneration
8. The Degree of Centralization
9. Line of Authority/Scalar Chain
10. Order
11. Equity
12. Stability of Tenure
13. Initiative
14. Team Spirit

AMERICAN AIRCRAFT

A3	Skywarrior	Douglas
A4	Skyhawk	Douglas
A6	Intruder	Grumman
A7	Corsair	Vought
A10	Thunderbolt	Fairchild Republic
A20	Boston/Havoc	Douglas
A24	Dauntless	Douglas
A25	Shrike	Curtiss
A26	Invader	Douglas
A28/29	Hudson	Lockheed
A35	Vengeance	Vultee
A36	Invader	North American
AC-130	Spectre/Stinger	Lockheed
AD	Skyraider	Douglas
AV-8B	Harrier	McDonnell Douglas
B1-B	B1 Bomber	Rockwell
B2	Stealth Bomber	Northrop
B4Y	Privateer	Consolidated
B17	Flying Fortress	Boeing
B18	Bolo	Douglas
B23	Dragon	Douglas
B24	Liberator	Consolidated
B25	Mitchell	North American
B26	Marauder	Martin
B29	Superfortress	Boeing
B36	Peacemaker	Convair

B45	Tornado	North American
B47	Stratojet	Boeing
B52	Stratofortress	Boeing
B58	Hustler	Convair
B66	Destroyer	Douglas
B70	Valkyrie	North American
C5	Galaxy	Lockheed
C9A	Nightingale	McDonnell Douglas
C17	Globemaster 3	Douglas
C46	Commando	Curtiss
C47	Dakota	Douglas
C53	Dakota	Douglas
C54	Skymaster	Douglas
C69	Constellation (military)	Lockheed
C97	Stratofreighter	Boeing
C113	Cargomaster	Douglas
C117	Skytrooper	Douglas
C118	Liftmaster	Douglas
C119	Flying Boxcar	Fairchild
C121	Constellation	Lockheed
C123	Provider	Fairchild
C124	Globemaster 2	Douglas
C125	Raider	Northrop
C130	Hercules	Lockheed
C131	Samaritan	Convair
C133	Cargomaster	Douglas
C135	Stratolifter	Boeing
C141	Starlifter	Lockheed

CV990	Coronado	Convair
DC4	Skymaster	Douglas
DC8	'Sunseeker'	Douglas
E2	Hawkeye	Grumman
E3	Sentry	Boeing
EF111	Raven	Grumman
EA3	Skywarrior	Douglas
EA6B	Prowler	Grumman
EA-18G	Growler	Boeing
F2A	Buffalo	Brewster
F4	Phantom	McDonnell Douglas
F4F	Wildcat	Grumman
F5	Freedom Fighter	Northrop
F6F	Hellcat	Grumman
F7C	Seahawk	Curtiss
F8F	Bearcat	Grumman
F9	Panther	Grumman
F9F	Sparrowhawk	Curtiss
F11(F)	Tiger	Grumman
F14	Tomcat	Grumman
F15	Eagle	McDonnell Douglas
F16	Fighting Falcon	General Dynamics
F17	Cobra	Northrop
F18	Hornet	McDonnell Douglas
F18E/F	Super Hornet	Boeing
F20	Tigershark	Northrop
F22	Raptor	Lockheed Martin/Boeing

F100	Super Sabre	North American
F35	Lightning II	Lockheed Martin
F101	Voodoo	McDonnell
F102	Delta Dagger	Convair
F104	Starfighter	Lockheed
F105	Thunderchief	Republic
F106	Delta Dart	Convair
F111	One eleven	General Dynamics
F117A	Nighthawk	Lockheed
KA3	Skywarrior	Douglas
KA6D	Intruder	Grumman
KC10	Extender	McDonnell Douglas
KC130	Hercules	Lockheed
KC135	Stratotanker	Boeing
L(x)49	Constellation	Lockheed
L1049	Super Constellation	Lockheed
L100	Hercules	Lockheed
L188	Electra	Lockheed
L1011	Tristar	Lockheed
P2	Neptune	Lockheed
P3	Orion	Lockheed
P4M	Mercator	Martin
P36	Hawk	Curtiss
P38	Lightning	Lockheed
P40	Warhawk	Curtiss
P43	Lancer	Republic
P47	Thunderbolt	Republic

P51	Mustang	North American
P61	Black Widow	Northrop
P70	Havoc	Douglas
P75	Eagle	General Motors
P80	Shooting Star	Lockheed
PBY	Catalina	Consolidated
PB4Y	Privateer	Consolidated
PT17	Stearman	Boeing
RB36	Peacemaker	Convair
S2	Tracker	Grummen
S3	Viking	Lockheed
SBC	Helldiver	Curtiss
SB2C	Helldiver	Curtiss
SC	Seahawk	Curtiss
SR71	Blackbird	Lockheed
TR1	Spyplane	Lockheed
U2	Spyplane	Lockheed
377	Stratocruiser	Boeing
737	'Baby Boeing'	Boeing
747	'Jumbo Jet'	Boeing
787	Dreamliner	Boeing

ANNIVERSARIES

1	Cotton, Paper
2	Paper, Cotton
3	**Leather**
4	Fruit, Flowers, Linen, Silk
5	**Wood**
6	**Iron**, Sugar (candy)
7	**Copper, Wool**, Brass
8	**Bronze**, Pottery
9	**Pottery**, Glass Crystal, Willow
10	**Tin**, Aluminium
11	**Steel**
12	Silk, Linen
13	**Lace**
14	**Ivory**, agate
15	**Crystal**, glass
20	**China**
25	**Silver**
30	**Pearl**
35	**Coral**, Jade
40	**Ruby**, Garnet
45	**Sapphire**, Tourmaline
50	**Gold**
55	**Emerald**
60	**Diamond**
70	Platinum

Latin names of anniversaries

25th	Quartocentennial
50th	Semicentennial
75th	Septiquinquennial
100th	Centennial
150th	Sesquincentennial
200th	Bicentennial
250th	Sesquibicentennial
300th	Tercentennial
400th	Quadricentennial
500th	Quincentennial

ATOMIC NUMBERS

No	Symbol	Element	Class
1	H	Hydrogen	non metal
2	He	Helium	noble gas
3	Li	Lithium	alkali metal
4	Be	Beryllium	alkaline earth
5	B	Boron	semi metal
6	C	Carbon	non metal
7	N	Nitrogen	non metal
8	O	Oxygen	non metal
9	F	Fluorine	halogen
10	Ne	Neon	noble gas
11	Na	Sodium	alkali metal
12	Mg	Magnesium	alkaline earth
13	Al	Aluminium	base metal
14	Si	Silicon	semi metal
15	P	Phosphorus	non metal
16	S	Sulphur	non metal
17	Cl	Chlorine	halogen
18	Ar	Argon	noble gas
19	K	Potassium	alkali metal
20	Ca	Calcium	alkaline earth
21	Sc	Scandium	transition metal
22	Ti	Titanium	transition metal
23	V	Vanadium	transition metal
24	Cr	Chromium	transition metal
25	Mn	Manganese	transition metal
26	Fe	Iron	transition metal

27	Co	Cobalt	transition metal
28	Ni	Nickel	transition metal
29	Cu	Copper	transition metal
30	Zn	Zinc	transition metal
31	Ga	Gallium	base metal
32	Ge	Germanium	semi metal
33	As	Arsenic	semi metal
34	Se	Selenium	non metal
35	Br	Bromine	halogen
36	Kr	Krypton	noble gas
37	Rb	Rubidium	alkali metal
38	Sr	Strontium	alkaline earth
39	Y	Yttrium	transition metal
40	Zr	Zirconium	transition metal
41	Nb	Niobium	transition metal
42	Mo	Molybdenum	transition metal
43	Tc	Technetium	transition metal
44	Ru	Ruthenium	transition metal
45	Rh	Rhodium	transition metal
46	Pd	Palladium	transition metal
47	Ag	Silver	transition metal
48	Cd	Cadmium	transition metal
49	In	Indium	base metal
50	Sn	Tin	base metal
51	Sb	Antimony	semi metal
52	Te	Tellurium	semi metal
53	I	Iodine	halogen
54	Xe	Xenon	noble gas
55	Cs	Caesium	alkali metal

56	Ba	Barium	alkaline earth
57	La	Lanthanum	lanthanide
58	Ce	Cerium	lanthanide
59	Pr	Praseodymium	lanthanide
60	Nd	Neodymium	lanthanide
61	Pm	Promethium	lanthanide
62	Sm	Samarium	lanthanide
63	Eu	Europium	lanthanide
64	Gd	Gadolinium	lanthanide
65	Tb	Terbium	lanthanide
66	Dy	Dysprosium	lanthanide
67	Ho	Holmium	lanthanide
68	Er	Erbium	lanthanide
69	Tm	Thulium	lanthanide
70	Yb	Ytterbium	lanthanide
71	Lu	Lutetium	lanthanide
72	Hf	Hafnium	transition metal
73	Ta	Tantalum	transition metal
74	W	Tungsten	transition metal
75	Re	Rhenium	transition metal
76	Os	Osmium	transition metal
77	Ir	Iridium	transition metal
78	Pt	Platinum	transition metal
79	Au	Gold	transition metal
80	Hg	Mercury	transition metal
81	Tl	Thallium	base metal
82	Pb	Lead	base metal
83	Bi	Bismuth	base metal
84	Po	Polonium	semi metal
85	At	Astatine	halogen

86	Em	Radon	noble gas
87	Fr	Francium	alkali metal
88	Ra	Radium	alkaline earth
89	Ac	Actinium	actinide
90	Th	Thorium	actinide
91	Pa	Protactinium	actinide
92	U	Uranium	actinide
93	Np	Neptunium	actinide
94	Pu	Plutonium	actinide
95	Am	Americium	actinide
96	Cm	Curium	actinide
97	Bk	Berkelium	actinide
98	Cf	Californium	actinide
99	Es	Einsteinium	actinide
100	Fm	Fermium	actinide
101	Mv	Mendelevium	actinide
102	No	Nobelium	actinide
103	Lr	Lawrencium	actinide
104	Rf	Rutherfordium	trans metal
105	Db	Dubnium	trans metal
106	Sg	Seaborgium	trans metal
107	Bh	Bohrium	trans metal
108	Hs	Hassium	trans metal
109	Mt	Meitnerium	trans metal
110	Ds	Darmstadtium	trans metal
111	Rg	Roentgenium	trans metal
112	Cn	Copernicium	trans metal
113	Nh	Nihonium	base metal
114	Fl	Flerovium	base metal
115	Mc	Moscovium	base metal

116	Lv	Livermarium	base metal
117	Ts	Tennessine	halogen
118	Og	Oganesson	noble gas

Magic Numbers

2	helium
8	oxygen
20	calcium
28	nickel
50	tin
82	lead
126	unbihexium (fictitious)

These relate to atoms with complete shells within the nucleus.

BINGO NUMBERS

1. Kelly's eye
2. A little duck, Baby's done it, Doctor Who
3. Dearie me, I'm free
4. The one next door, On the floor
5. Man alive, Jack's alive
6. Tom Mix, Chopsticks
7. Lucky seven, Gawd's in 'eaven
8. Garden gate, Golden Gate
9. Doctor's orders
10. Downing Street
11. Legs eleven
12. One Dozen, Monkey's cousin
13. Unlucky for some
14. Valentine's day
15. Rugby team
16. Sweet sixteen and never been kissed
17. Often been kissed, Old Ireland
18. Key of the door, Now you can vote
19. Goodbye teens
20. One score, Getting plenty
21. Royal salute
22. Two little ducks
23. The Lord's my Shepherd
24. Two dozen
25. Duck and dive
26. Bed and breakfast, Half a crown
27. Little duck with a crutch

28. In a state
29. In your prime
30. Burlington Bertie, Dirty Gertie
31. Get up and run
32. Buckle my shoe
33. Dirty knees, All the feathers
34. Ask for more
35. Jump and jive
36. Three dozen
37. A flea in heaven
38. Christmas cake
39. Those famous steps
40. Two score, Life begins at
41. Time for fun
42. That famous street in Manhattan
43. Down on your knees
44. Droopy draws
45. Halfway house, Halfway there
46. Up to tricks
47. Four and seven
48. Four dozen
49. PC, Nick nick
50. Bulls eye, Bung hole
51. I love my Mum
52. Weeks in a year
53. Stuck in a tree
54. House of Bamboo
55. Snakes alive
56. Shott's bus
57. Heinz varieties

58. Make them wait
59. Brighton line
60. Three score
61. Baker's bun
62. Tickety boo
63. Tickle me
64. The Beatles Number
65. Old age pension
66. Clickety click
67. Made in heaven
68. Saving grace
69. The same both ways, your place or mine?
70. Three score and ten
71. Bang on the drum
72. A crutch and a duck
73. Crutch with a flea
74. Candy store
75. Big Daddy
76. Was she worth it?
77. Sunset Strip
78. Heaven's gate
79. One more time
80. Gandhi's breakfast
81. Fat lady and a little wee
82. Fat lady with a duck
83. Fat lady with a flea
84. Seven dozen
85. Stayin' alive
86. Between the sticks

87. Torquay in Devon
88. Two fat ladies, Wobbly wobbly
89. Nearly there
90. Top of the shop, Top of the House

COMPARATIVE NUMBERS, ASCII CODE AND BASES

Arabic	Roman	Binary
1	I	1
2	II	10
3	III	11
4	IV	100
5	V	101
6	VI	110
7	VII	111
8	VIII	1 000
9	IV	1 001
10	X	1 010
11	XI	1 011
12	XII	1 100
13	XIII	1 101
14	XIV	1 110
15	XV	1 111
16	XVI	10 000
17	XVII	10 001
18	XVIII	10 010
19	XIX	10 011
20	XX	10 100
21	XXI	10 101
30	XXX	11 110
32	XXXII	100 000

40	XL	101 000
50	L	110 010
60	LX	111 100
64	LXIV	1 000 000
70	LXX	1 000 110
80	LXXX	1 010 000
90	XC	1 011 010
100	C	1 100 100
128	CXXVIII	10 000 000
200	CC	11 001 000
256	CCLVI	100 000 000
300	CCC or B	100 101 100
400	CD	110 010 000
500	D	111 110 100
512	DXII	1 000 000 00
1000	M	1 111 101 000
1024	MXXIV	10 000 000 000
2000	MM	11 111 010 000
2048	MMXLVIII	100 000 000 000
3000	MMM	101 110 111 000
4000	MV	111 110 100 000
4096		1 000 000 000 000
5000	V	1 001 110 001 000
8,192		10 000 000 000 000
10,000	X	10 011 100 010 000
16,384		100 000 000 000 000
20,000	XX	100 111 000 100 000
32,768		1 000 000 000 000 000
50,000	L	1 100 001 101 010 000

65,536............	....	10 000 000 000 000 000
100,000...........	C........................	11 000 011 010 100 000
131,072...........	....	100 000 000 000 000 000
262,144...........	....	1000 000 000 000 000 000
500,000...........	D........................	1 111 010 000 100 100 000
524,288...........	....	10 000 000 000 000 000 000
1,000,000.........	M........................	11 110 100 001 001 000 000

ASCII code

SPACE	032	!	033
"	034	#	035
$	036	%	037
&	038	'	039
(	040	)	041
*	042	+	043
,	044	_	045
.	046	/	047
0	048	1	049
2	050	3	051
4	052	5	053
6	054	7	055
8	056	9	057
:	058	;	059
<	060	=	061
>	062	?	063
@	064	A	065
B	066	C	067
D	068	E	069

F	070	G	071
H	072	I	073
J	074	K	075
L	076	M	077
N	078	O	079
P	080	Q	081
R	082	S	083
T	084	U	085
V	086	W	087
X	088	Y	089
Z	090	[	091
\	092	]	093
^	094	backspace	095
space	096	a	097
b	098	c	099
d	100	e	101
f	102	g	103
h	104	i	105
j	106	k	107
l	108	m	109
n	110	o	111
p	112	q	113
r	114	s	115
t	116	u	117
v	118	w	119
x	120	y	121
z	122	;	123
<	124	=	125
>	126	delete	127

Number bases

0	nunnary
1	urinary
2	binary
3	ternary
4	quaternary
8	octal
10	decimal
16	hexadecimal
32	base 32
64	base 64

CONSTANTS AND PI ETC

Pi = 3.141 592 653 5…. etc
Mnemonics for remembering pi are numerous here are a few:

How I wish I could recollect pi easily today!
May I have a large container of coffee, cream and sugar?
How I need a drink, alcoholic in nature, after the tough chapters involving quantum mechanics!

Euler's Number (e) = 2.718 281 828 459…..

Pythagoras' Constant (square root of 2) = 1.414 213 562 373….

The Golden Ratio = 1.618 033 988 749….

Conway's Constant = 1.303 57….

Khinchin's Constant = 2.685 452 001….

Glaiser – Kinkelin Constant = 1.282 427 129 1….

Theodorus' Constant (square root of 3) = 1.732 050 807 568….

Light speed (c) = 299 792 458 meters/second

Gravitational Constant (G) = 6.674 x 10(-11) m3/kg/sec/sec

Planck length = 1.616 229 x 10(-35) m

Planck mass = 2.176 470 x 10(-8) kg

Planck time = 5.391 16 x 10(-44) sec

Planck charge = 1.875 545 956 x 10(-18) coulombs

Planck temperature = 1.416 808 x 10(32) deg K

Cook's Constant = whatever you want it to be!

CONTAINERS

Champagne bottles

- 2 bottles.......................Magnum
- 4 bottles......................Jeroboam
- 6 bottles....................Rehoboam
- 8 bottles ....Methuselah
- 12 bottles..................Salmanazar
- 16 bottles ....Balthazar
- 20 bottles .... Nebuchadzezzar

One may journey responsibly meaning Sally backs Neb

One nip equals a quarter bottle
A baby equals one eighth of a bottle

Beer barrels

- Firkin...............................9 gallons
- Kilderkin.......................18 gallons
- Barrell...........................36 gallons
- Hogshead.....................54 gallons
- Butt.............................108 gallons
- Tun.............................216 gallons

Fern killed Barry Hoggrill's bush tucker

Wine barrels

- Anker............................10 gallons
- Hogshead......................63 gallons
- Puncheon......................84 gallons
- Pipe..............................126 gallons
- Tun..............................252 gallons
- Butt (sherry)................110 gallons

DEFINITIONS

Apocalyptic Number

        666   (Revelation 8:18)

The Golden Number (Book of Common Prayer)

A number between 1 and 19, used to indicate the position of any year in the Metonic cycle, calculated as the remainder when 1 is added to the given year and the sum is divided by 19. If the remainder is zero the number is 19.

> Add one to the Year of Our Lord, and then divide by 19;
> The remainder, if any, is the Golden Number
> But if nothing remaineth, then 19 is the Golden Number

Perfect numbers

> An integer that is the sum of all its possible factors
> The first four are:

6	(1+2+3)
28	(1+2+4+7+14)
496	
8128	

Morse numbers

1	.＿＿＿＿
2	..＿＿＿
3	...＿＿
4	....＿
5	.....
6	＿....
7	＿＿...
8	＿＿＿..
9	＿＿＿＿.
10	＿＿＿＿＿

Prime numbers

An integer that is only divisible by itself and one:

2, 3, 5, 7, 11, 13, 17, 19, 23, 29, 31, 37, 41, 43, 47, 53, 59, 61, 67, 71, 73, 79, 83, 89, 97, 101, 103, 107, 109, 113, 127, 131, 137, 139, 149, 151, 157, 163, 167, 173, 179, 181, 191, 193, 197, 199, 211, 223, 227, 229, 233, 239, 241, 251, 257, 263, 269, 271, 277, 281, 283, 293, 307, 311, 313, 317, 331, 337, 347, 349, 353, 359, 367, 373, 379, 383, 389, 397, 401, 409, 419, 421, 431, 433, 439, 443, 449, 457, 461, 463, 467, 479, 487, 491, 499, 503 etc.

Composite Numbers

Any whole number which is not a prime:

4, 6, 8, 9, 10, 12, 14, 15, 16, 18, 20, 21, 22, 24, 25 etc....

Fibonacci sequence

Series where each number is the sum of the two numbers preceding it:

$$0 + 1 = 1$$
$$1 + 1 = 2$$
$$2 + 1 = 3$$
$$3 + 2 = 5$$
$$5 + 3 = 8$$
$$8 + 5 = 13$$
$$13 + 8 = 21$$
$$21 + 13 = 34$$
$$34 + 21 = 55$$
$$55 + 34 = 89$$
$$89 + 55 = 144$$
$$144 + 89 = 233$$
$$233 + 144 = 377$$
$$377 + 233 = 610$$
$$610 + 377 = 987$$
$$987 + 610 = 1597 \quad \text{etc.}$$

Squares and Cubes

A square is the product of a number multiplied by itself.
A cube is the product of a number multiplied by itself twice.

Number	Square	Cube
1	1	1
2	4	8
3	9	27
4	16	64
5	25	125

6	36	216
7	49	343
8	64	512
9	81	729
10	100	1000
11	121	1331
12	144	1728
13	169	2197
14	196	2744
15	225	3375
16	256	4096
17	289	4913
18	324	5832
19	361	6859
20	400	8000

Mach Number
The ratio of the speed of a body in a particular medium to the speed of sound in that medium.

i.e Mach 1.0 equals the speed of sound

Froude Number
Used in fluid mechanics and modelling to describe the nature of the flow. The Critical Froude Number = 1.0

Reynolds Number
Used in fluid mechanics to show the nature of flow i.e. whether turbulent or laminar.

Harshad Number
An integer divisible by the sum of its digits in a given base:

1, 2, 3, 4, 5, 6, 7, 8, 9, 10, 12, 18, 20, 21, 24, 27, 30, 36, etc…

Strobogramatic Number
A number which is the same when turned through 180 degrees e.g:

69, 96, 101, 609, 9006, 91116

Smith Numbers
A composite number for which, in a given base, the sum of its digits is equal to the sum of the digits in its prime factorization e.g:

4 22 27 58 85 94 121 166 202 etc…

Pronic numbers
Numbers which are the product of two consecutive integers e.g:

0, 2, 6, 12, 20, 30, 42, 56, 72, 90, 110, 132, 156, 182, etc…

Mersenne Numbers
Numbers in a series of 2 to the power 'n' minus 1, i.e:

0, 1, 3, 7, 15, 31, 63, 127, 255, 511, 1023, 2047, etc.

Trancendental Numbers
Numbers which cannot be expressed using a finite number of arithmetical expressions.

Triangular Numbers
Count the number of objects that can be arranged in the form of an equilateral triangle, e.g:

0, 1, 3, 6, 10, 15, 21, 28, 36, 45, 55, 66, 78, 91, 105, 120

Rational Numbers
Numbers which can be expressed in the form: **a/b**
where a and b are integers and b is not 0, e.g:

2, 25, 7/3, 256/13, 13/256

Irrational Numbers
Real numbers that cannot be expressed as a ratio of two integers, e.g:

Pi

Real Numbers
All rational and irrational numbers, i.e. all integers and fractions and decimals and combinations of these but not complex numbers:

21, 1.35, 999.99, 3.146235749......, 16¾

Imaginary numbers
Numbers which are based on the square root of -1.

Complex Numbers
Numbers which are partly unreal in the form:

a + bi

where a and b are real but i is the square root of -1

Cardinal Numbers
Numbers denoting quantity rather than position such as:

6, 45, 89, 245, 12,456

Ordinal Numbers
Numbers denoting relative position in a sequence such as:

First, second, eighth, twenty first

Integers
Whole numbers such as:

1, 5, 10, 56 or 1478

Untouchable numbers

Positive integers that cannot be expressed as the sum of all the proper divisors of any positive integer (including the untouchable number itself) e.g:

2, 5, 52, 88, 96, 120, 124, 146, 162, 188, 206, 210, 216, etc…

Happy Numbers

Positive integers which, when the sum of the squares of the digits are repetitively added, it will revert to 1, e.g:

1, 7, 10, 13, 19, 23, 28, 31, 32, 44, 49, 68, 70, 79, 82, 86

(An unhappy or 'sad' number does not revert to 1.)

Sphenic numbers

Are the result of the product pqr where p, q, and r are three distinct prime numbers e.g:

30, 42, 66, 70, 78, 102, 105, 110, 114, 130, 138, 154, etc…

Polygonal Numbers

Numbers which can be arranged as dots in the shape of a regular polygon, for instance:

Triangular: 1, 3, 6, 10, 15, etc…
Square: 1, 4, 9, 16, 25, etc…
Pentagonal: 1, 5, 12, 22, 35, etc…
Hexagonal: 1, 6, 15, 28, 45, etc…

UN Numbers

Used in the Transport of Dangerous Goods

3001 to 3100

Numbers Game
A form of illegal lottery played in the USA

Memorable Phone Numbers
These are desirable phone numbers which are easy to remember because they involve repetition or a sequence, e.g:

Gold numbers
01827 000 555 - a triple-triple number
01827 123 4567 – a sequential number
01827 256 256 – a repetitive number
0187 94 94 94 - another repetitive number

Silver numbers
01827 11 33 66 - double-double-double
01827 111 97 97- triple-double

Bronze numbers
01827 432 333 - a single triple
01827 007 007 - obvious

ISBN
International Standard Book Numbers are a system which even Wikipedia struggles to explain hence a lack of detail in this book. It originally consisted of ten digits and was expanded to thirteen in 2007:

Prefix or EAN – 3 digits
Registration group – 2 digits
Registration element (publisher) – 4 digits
Publication element (title) – 3 digits
Check digit – 1 digit
Total – 13 digits

However the three initial digits are complicated as they can sometimes be a country code, a language code or just 978 (and now 979). Some numbers are reserved for applications in the US and for sheet music.

DDC

The Dewey Decimal Classification is a logical system of classification using only decimal numbers. The first order is based on:

000 – Computer science, info & general
100 – Philosophy and psychology
200 – Religion
300 – Social sciences
400 – Language
500 – Pure Science
600 – Technology
700 – Arts & recreation
800 – Literature
900 – History & geography

For example:

500	Science
510	Mathematics
516	Geometry
516.3	Analytical geometry
516.37	Metric differential geometry
516.375	Finster geometry

It suffers from inflexibility and overloading of certain sections.

LCCN

The American Library of Congress Control Number is another complex system which originated in 1898 and was amended in 2001.

Pre 2000 system
Alpha code – 3 digits
Year code – 2 digits
Serial number – 6 digits
Supplement/ rev date - + digits

Post 2000 system
Alpha code – 2 digits
Year code – 4 digits
Serial number – 6 digits

EDUCATION IN MATHEMATICS

All mathematics is based on numbers and is learned in a sequential manner getting harder with each stage; this table is somewhat subjective and serves only as rough guidance for when subjects are introduced to students.

Year (UK)	Age	K-12	Subjects introduced
Reception	3-4	PK	Money, size, patterns, shapes
Year 1	5	K	Counting, time, simple addition
Year 2	6	K1	Addition, subtraction, estimation
Year 3	7	K2	Multiplication, division, fractions, measurement, 2d shapes, units
Year 4	8	K3	Simple graphs, 3d shapes, geometry
Year 5	9	K4	Decimals, probabilities
Year 6	10	K5	Quadrilaterals, percentages, number theory, coordinates
Year 7	11	K6	Algebra, roots, powers, rates of change
Year 8	12	K7	Geometry, measures, quadratic equations
Year 9	13	K8	Probability, ratios, complex graphs

Year 10	14	K9	Conversions, statistics, factorizing, polynomials
Year 11	15	K10	Vectors, arrays, logarithms, area and volume
Year 12	16	K11	Trigonometry, polynomials, pure maths, limits
Year 13	17	K12	Calculus, mechanics non-real numbers, series

E NUMBERS

This list is an extract containing the commoner substances only.

E100-E199 (colours)

E100	**Curcumin (from turmeric)**	Yellow-orange
E101	**Riboflavin (Vitamin B2)**, formerly called lactoflavin	Yellow-orange
E102	**Tartrazine** (FD&C Yellow 5)	Lemon yellow
E104	Quinoline Yellow WS	Dull/green yellow
E110	**Sunset Yellow** FCF (Orange Yellow S, FD&C Yellow 6)	Yellow-orange
E111	Orange GGN	Orange
E120	**Cochineal**, Carminic acid, Carmine (Natural Red 4)	Crimson
E121	Citrus Red 2	Dark red
E123	Amaranth	Dark red
E128	Red 2G	Red
E131	Patent Blue V	Dark blue
E132	Indigo carmine	Indigo
E133	Brilliant Blue FCF	Reddish blue
E140	**Chlorophylls**	Green
E142	Green S	Green
E143	Fast Green FCF (FD&C Green 3)	Sea green
E150	**Caramels**	Brown
E151	Black PNs	Black
E152	**Carbon black (hydrocarbon)**	Black
E153	Vegetable carbon	Black
E154	Brown FK (kipper brown)	Brown

E155	Brown HT (chocolate brown HT)	Brown
E160b	Annatto, bixin, norbixin	Orange
E162	Beetroot Red, Betanin	Red
E164	**Saffron**	Orange-red
E170	Calcium carbonate, **Chalk**	White
E171	**Titanium dioxide**	White
E172	Iron oxides and hydroxides	Brown
E173	**Aluminium**	Silver to grey
E174	**Silver**	Silver
E175	**Gold**	Gold
E180	Pigment Rubine, Lithol Rubine BK	Red
E181	**Tannin**	Brown

E200–E299 (preservatives)

E200	Sorbic acid	preservative
E201-203	Sorbates	preservative
E210	Benzoic acid	preservative
E211-213	Benzoates	preservative
E220	**Sulphur dioxide**	preservative
E221	Sodium sulphite	preservative
E222	Sodium bisulphite	preservative
E223	**Sodium metabisulphite**	preservative
E224	Potassium metabisulphite	preservative
E225	Potassium sulphite	preservative
E238	Calcium formate	preservative
E240	**Formaldehyde**	preservative
E249	Potassium nitrite	preservative
E250	Sodium nitrite	preservative
E251	Sodium nitrate (**Chile saltpeter**)	preservative
E252	Potassium nitrate (**Saltpetre**)	preservative

E260	**Acetic acid (vinegar)**	acidity regulator
E261	Potassium acetate	acidity regulator
E263	Calcium acetate	acidity regulator
E264	Ammonium acetate	preservative
E270	**Lactic acid**	antioxidant
E280	Propionic acid	preservative
E281-283	Propionates	preservative
E284	Boric acid	preservative
E285	Sodium tetraborate (**borax**)	preservative
E290	**Carbon dioxide**	acidity regulator
E296	Malic acid	acidity regulator
E297	Fumaric acid	acidity regulator

E300–E399 (antioxidants, acidity regulators)

E300	**Ascorbic acid (Vitamin C)**	antioxidant
E301-303	Ascorbates	antioxidant
E304	Fatty acid esters of ascorbic acid	antioxidant
E305	Ascorbyl stearate	antioxidant
E306	Tocopherols (**Vitamin E, natural**)	antioxidant
E314	Guaiac resin	antioxidant
E322	Lecithins	emulsifier
E325-329	Lactates	acidity regulator
E330	**Citric acid**	acid regulator
E331-333	Citrates	acidity regulator
E334	**Tartaric acid**	acid
E335	Sodium tartrates	acidity regulator
E336	Potassium tartrates (**cream of tartar**)	acidity regulator
E337	Sodium potassium tartrate (**Rochelle salt**)	acidity regulator
E338	Orthophosphoric acid	acid

E339-343	Phosphates	antioxidant
E344	Lecitin citrate	acidity regulator
E345	Magnesium citrate	acidity regulator
E349-352	Malates	acidity regulator
E353	Metatartaric acid	emulsifier
E354	Calcium tartrate	emulsifier
E355	Adipic acid	acidity regulator
E356-359	Adipates	acidity regulator
E363	Succinic acid	acidity regulator
E365-368	Fumarates	acidity regulator
E385	Calcium disodium **EDTA**	sequestrant
E387	Oxystearin	stabiliser
E391	Phytic acid	
E392	Extracts of rosemary	

E400–E499 (thickeners, stabilisers, emulsifiers)

E400	Alginic acid	emulsifier
E401-405	Alginates	emulsifier
E406	**Agar**	stabiliser
E407a	Processed eucheuma seaweed	emulsifier
E408	**Bakers' yeast** glycan	
E410	Locust bean gum (**Carob gum**)	emulsifier
E411	**Oat gum**	stabiliser
E412	Guar gum	stabiliser
E413	Tragacanth	emulsifier
E414	Acacia gum (**gum arabic**)	emulsifier
E415-419	Gums	stabiliser
E420	**Sorbitols**	humectant
E421	Mannitol	sweetener
E422	**Glycerol**	sweetener

E424	Curdlan	gelling agent
E425	Konjac	emulsifier
E427	**Cassia gum**	
E429	**Peptones**	
E440	**Pectins**	emulsifier
E441	**Gelatine**	gelling agent
E460	**Cellulose**	emulsifier
E461-468	Celluloses	emulsifier
E490	**Propane**-1,2-diol	

E500–E599 (acidity regulators, anti-caking agents)

E500-505	Carbonates	raising agent
E507	**Hydrochloric acid**	acid
E508	Potassium chloride	seasoning
E509	Calcium chloride	firming agent
E510-512	Chlorides	improving agent
E513	**Sulphuric acid**	acid
E514-523	Sulphates	acid
E518	Magnesium sulphate (**Epsom salts**)	firming agent
E524-528	Hydroxides	acidity regulator
E529-530	Oxides	improving agent
E535-536	Ferrocyanides	anti-caking agent
E542	Bone phosphate	anti-caking agent
E543-545	Polyphosphates	emulsifier
E550	Sodium Silicates	anti-caking agent
E551	Silicon dioxide (**Silica**)	anti-caking agent
E552	Calcium silicate	anti-caking agent
E553a	Magnesium silicates	anti-caking agent
E553b	**Talc**	anti-caking agent
E554-560	Silicates	anti-caking agent

E558	**Bentonite**	anti-caking agent
E559	Aluminium silicate (**Kaolin**)	anti-caking agent
E560	Potassium silicate	anti-caking agent
E561	**Vermiculite**	
E562	Sepiolite	
E563	Sepiolitic clay	
E570	**Fatty acids**	anti-caking agent
E572	Magnesium stearate, calcium stearate	anti-caking agent
E574	Gluconic acid	acidity regulator
E575	Glucono delta-lactone	sequestrant
E576-580	Gluconates	sequestrant
E585	Ferrous lactate	food colouring
E599	**Perlite**	

E600–E699 (flavour enhancers)

E620	Glutamic acid	flavour enhancer
E621	**Monosodium glutamate (MSG)**	flavour enhancer
E626	Guanylic acid	flavour enhancer
E630	Inosinic acid	flavour enhancer
E631	Disodium inosinate	flavour enhancer
E640	Glycine and its sodium salt	flavour enhancer
E650	Zinc acetate	flavour enhancer

E700–E799 (antibiotics)

E701	**Tetracyclines**	antibiotic
E702	Chlortetracycline	antibiotic
E703	Oxytetracycline	antibiotic
E704	Oleandomycin	antibiotic
E705-708	**Penicillins**	antibiotic
E710	Spiramycins	antibiotic

| E711 | Virginiamycins | antibiotic |
| E712 | Flavomycin | antibiotic |

E900–E999 (glazing agents, gases and sweeteners)

E901	**Beeswax**, white and yellow	glazing agent
E902	Candelilla wax	glazing agent
E903	Carnauba wax	glazing agent
E904	**Shellac**	glazing agent
E905	**Paraffins**	
E905a	Mineral oil	anti-foaming
E905c	Petroleum waxes	glazing agent
E906	Gum benzoic	flavour enhancer
E907	Crystalline wax	glazing agent
E908	Rice bran wax	glazing agent
E909	Spermaceti wax	glazing agent
E910	Wax esters	glazing agent
E911	Methyl esters of fatty acids	glazing agent
E912	Montanic acid esters	glazing agent
E913	**Lanolin**, sheep wool grease	glazing agent
E916	Calcium iodate	
E917	Potassium iodate	
E918	Nitrogen oxides	
E919	Nitrosyl chloride	
E922	Potassium persulphate	improving agent
E923	Ammonium persulphate	improving agent
E924	Potassium bromate	improving agent
E924b	Calcium bromate	improving agent
E925	**Chlorine**	**bleach**
E926	**Chlorine dioxide**	bleach
E927a	Azodicarbonamide	improving agent

E927b	Carbamide (**urea**)	improving agent
E928-930	Peroxides	bleach
E938	**Argon**	packaging gas
E939	**Helium**	packaging gas
E941	**Nitrogen (packaging gas)**	propellant
E942	**Nitrous oxide**	propellant
E943a	**Butane**	propellant
E943b	Isobutane	propellant
E944	**Propane**	propellant
E948	**Oxygen**	packaging gas
E949	**Hydrogen**	packaging gas
E951	**Aspartame**	sweetener
E952	Cyclamic acid and Cyclamate	sweetener
E954	**Saccharins**	sweetener
E955	Sucralose	sweetener
E956	Alitame	sweetener
E957	Thaumatin	flavour enhancer
E958	Glycyrrhizin	flavour enhancer
E961	Neotame	sweetener
E965	Maltitol	humectant
E966	Lactitol	sweetener
E967	Xylitol	sweetener
E968	Erythritol	sweetener
E999	Quillaia extract	foaming agent

E1000–E1599 (additional additives)

E1000	Cholic acid	emulsifier
E1001	Choline salts	emulsifier
E1100	**Amylase**	stabiliser
E1102	Glucose oxidase	antioxidant

E1104	**Lipases**	
E1105	Lysozyme	preservative
E1204	Pullulan	
E1400	**Dextrins**	thickening agent
E1403	Bleached starch	thickening agent
E1411	Distarch glycerol	emulsifier
E1430	Distarch glycerine	thickening agent
E1501	Benzylated hydrocarbons	
E1502	**Butane**-1, 3-diol	
E1503	**Castor oil**	resolving agent
E1504	Ethyl acetate	flavour solvent
E1505	Triethyl citrate	foam stabiliser
E1510	**Ethanol**	
E1519	**Benzyl alcohol**	
E1520	Propylene glycol	humectant

FOREIGN NUMBERS

	Spanish	German	French	Italian
1	Uno	Ein	Un	Uno
2	Dos	Zwei	Deux	Due
3	Tres	Drei	Trois	Tre
4	Cuatro	Vier	Quatre	Quattro
5	Cinco	Funf	Cinq	Cinque
6	Seis	Sechs	Six	Sei
7	Siete	Sieben	Sept	Sette
8	Ocho	Acht	Huit	Otto
9	Nueve	Nuen	Neuf	Nove
10	Diez	Zehn	Dix	Dieci
11	Once	Elf	Onze	Undici
12	Doce	Zwolf	Douze	Dodici
13	Trece	Dreizehn	Treize	Tredici
14	Catorce	Vierzehn	Quatorze	Quattordici
15	Quince	Funfzehn	Quinze	Quindici
16	Dieciseis	Sechszehn	Soixante	Seidici
17	Diecisiete	Siebzehn	Dix-sept	Diciasette
18	Dieciocho	Actzehn	Dix-huit	Diciotto
19	Diecinueve	Neunzehn	Dix-neuf	Dicianove
20	Veinte	Zwanzig	Vingt	Venti
21	Veintiun/o/a	Einundzwanzig	Vingt-et-un	Venuno
30	Treinta	Dreissig	Trente	Trenta
40	Cuarenta	Vierzig	Quarante	Quaranta
50	Cincuenta	Funfzig	Cinquante	Cinquanta
60	Sesenta	Sechzig	Soixante	Sessanta
70	Setenta	Siebzig	Soixante-dix	Settanta

80	Ochenta	Achtzig	Quatre-vingt	Ottanta
90	Noventa	Nuenzig	Quatre-vingt-dix	Novanta
100	Cien or ciento	Hundert	Cent	Cento
200	Doscientos/as	Zwei hundert	Deux cent	Due cento
300	Trescientos/as	Drei hundert	Trois cent	Tre cento
500	Quincientos/as	Funf hundert	Cinque cent	Cinque cento
1000	Mil	Tausend	Mille	Mille

FREQUENCIES AND TIME PERIODS

Frequency

Hourly	Once an hour
Semidiurnal	Twice a day
Daily	Once a day
Diurnal	Daily
Biweekly	Twice a week or every 2 weeks
Weekly	Once a week
Bimonthly	Twice a month or every 2 months
Monthly	Once a month
Biannual	Twice a year or every 2 years
Semiannual	Every six months
Annually	Once a year
Perennial	Every year
Blue moon*	Rarely, roughly once a year
Biennial	Once in 2 years
Triennial	Every 3 years
Quadrennial	Every 4 years
Sexennial	Every 6 years
Septennial	Every 7 years
Octennial	Every 8 years
Novennial	Every 9 years
Decennial	Once a decade / in 10 years
Undecennial	Every 11 years
Duodecennial	Once in 12 years
Quindecennial	Every 15 years

Vicennial	Every 20 years
Tricennial	Every 30 years
Semi-centennial	Every 50 years
Centennial	Once a century / in 100 years
Sesquicentennial	Every 150 years
Bicentenial	Every 200 years
Quadricentennial	Every 400 years
Quincentennial	Every 500 years

*relates to there being two new moons in the same callendar month

Time Periods

Millisecond	1/1000 of a second
Second	basic unit
Minute	60 seconds
Hour	60 minutes
Day	24 hours
Week	7 days
Fortnight	14 days
Month	28 – 31 days
Quarter	12 weeks
Season	approx 3 months
Trimester	three months
Semester	six months
Year	365 days (actually 365.25)
Leap year	366 days
Year	52 weeks or 12 months
Olympiad	4 years
Pentad	5 years

Lustrum	5 years
Decade	10 years
Indiction	15 years
'Preston Guild'	every 20 years
Half century	50 years
Century	100 years
Millennium	1000 years
Aeon	Very long time or eternity

GAMES AND SPORTS

The tile values in Scabble*

Letter	No of tiles	Value	Letter	No of tiles	Value
A	9	1	**B**	2	3
C	2	3	**D**	4	2
E	12	1	**F**	2	4
G	3	2	**H**	2	4
I	9	1	**J**	1	8
K	1	5	**L**	4	1
M	2	3	**N**	6	1
O	8	1	**P**	2	3
Q	1	10	**R**	6	1
S	4	1	**T**	6	1
U	4	1	**V**	2	4
W	2	4	**X**	1	8
Y	2	4	**Z**	1	10
BLANK	2	0			

The Scrabble* squares

(15x15 makes 225 altogether)

8 triple word spaces

12 triple letter spaces

17 double word spaces

24 double letter spaces

(61 bonus spaces altogether)

* Scrabble is a registered trademark of Spears Games

Snooker Balls

- Black   (7)
- Pink   (6)
- Blue   (5)
- Brown   (4)
- Green   (3)
- Yellow (2)
- Reds   (1)
- White   (-4)

Tarot Cards - The 22 'Trump' cards of the Major Arcana

1. Juggler (Magician)
2. Female Pope (Papess)
3. Empress
4. Emperor
5. Pope
6. Lovers
7. Chariot
8. Justice
9. Hermit
10. Wheel of Fortune

11. Strength (Fortitude)   12. Hanged Man
13. Death   14. Termperence
15. The Devil   16. Tower (struck by lightning)
17. Star   18. Moon
19. Sun   20. Last (Day of) Judgement
21. World (Universe)   No number   Fool

Plus 56 cards in the Minor Arcana:

Ace, 1, 2, 3, 4, 5, 6, 7, 8, 9, 10, Page, Knight, Queen, King in each of four suits:

Staves (wands) equivalent to:	Clubs
Cups	Hearts
Swords	Spades
Coins	Diamonds

Darts

Order of the numbers on a 'London Board':

20 1 18 4 13 6 10 15 2 17
 3 19 7 16 8 11 14   9 12 5

Order of the numbers on an 'East End' or 'Fives Board':

20 5 15 10 20 5
15 10 20   5 15 10

Order of the numbers on a 'Manchester Board':

4 20 1 16 6 17 8 12 9 14
5 19 2 15 3 18 7 11 10 13

The is no treble ring on a Manchester board just a double ring.

Nine dart finishes (on a 'London Board')

$$60 + 60 + 60 = 180 \quad (180)$$
$$60 + 60 + 60 = 180 \quad (360)$$
$$60 \quad\quad = 60 \quad (420)$$

treble 19 plus double 12
or treble 17 plus double 15 = 81 (501)
or treble 15 plus double 18

Sports Teams

2	Croquet
4	Polo and Curling
5	Basketball
6	Volleyball and Ice Hockey
7	Water Polo and Netball Handball and Kabaddi
8	Tug of War
9	Baseball and Rounders Softball and Kho-Kho
10	Lacrosse (Men)
11	Field Hockey and American Football Cricket and Soccer

12   Lacrosse (Women) and Shinty
        Canadian football

13   Rugby League

15   Rugby Union and Gaelic Football
        Hurling

18   Australian Rules Football

GASMARKS

Gasmark	Deg C	Deg F
1/2 (slow)	120	250
1	140	275
2	150	300
3	170	325
4 (moderate)	180	350
5	190	375
6 (hot)	200	400
7	220	425
8 (very hot)	230	450
9	260	500

Radioactive Isotopes

Isotope	Radiation	Half life
Carbon 14	beta	5,600 years
Phosphorus 32	beta	14.3 days
Cobalt 60	beta, gamma	5.3 years
Strontium 90	beta	28 years
Iodine 131	beta, gamma	8 days
Caesium 137	beta, gamma	30 years
Radium 226	alpha, gamma	1,620 years
Uranium 235	alpha	710 m years
Uranium 238	alpha	4,510 m years
Plutonium 239	alpha	24,400 years

GEOMETRY

Polygons

Sides	name	internal Angles	sum of angles
3	triangle	60 deg	180 deg
4	square	90	360
5	pentagon	108	540
6	hexagon	120	720
7	heptagon	128.6	900
8	octagon	135	1080
9	nonagon	140	1260
10	decagon	144	1440
11	undecagon	147.3	1620
12	dodecagon	150	1800

Triangles

Equilateral	all sides equal
Isosceles	two sides (and angles) equal
Scalene	all sides (and angles) different
Right angle	contains one angle of 90 deg

Quadrilaterals

Square	all sides equal and all angles 90 deg
Rectangle	opposite sides equal and all angles 90 deg
Rhombus	all sides equal but no right angles

Parallelogram	opposite side equal and parallel
Trapezium	one pair of opposite side parallel
Kite	adjacent sides equal and diagonals intersect at 90 deg

GEOGRAPHIC COORDINATES

Precise location on the Earth's surface is determined by a system of coordinates based on the Equator for 'vertical' measurement and the Prime Meridian (at Greenwich Observatory) for 'horizontal' location.

Latitude
Measurement is defined in degrees, minutes and seconds with the equator as 0 degrees, the North Pole as 90 deg N and the South Pole as 90 deg S. The Arctic Circle is at 66deg 33min N and the Tropic of Cancer at 23deg 27min N. The Tropic of Capricorn is at 23deg 27min S and the Antarctic Circle at 66deg 33min S.

Longitude
Measurement, in degrees, minutes and seconds, was eastwards from the Prime Meridian at Greenwich until the circle was complete at 360 deg. Now it is more normally seen as plus degrees to the east and minus degrees to the west. The Prime Meridian was internationally agreed in 1884. Each line in the 'vertical' plane is a 'meridian'. The International Dateline serves as a proxy for the 180deg meridian and is shifted to preserve the integrity of Pacific nations.

Digital Degrees (DMS)
Both Latitude and longitude are now normally shown in digital degrees rather than degrees, minutes and seconds; this aids calculation as base 10 is easier to work with than base 60 (sexagesimal). A place in New York state with the latitude and longitude: 43°2'27" N and 77°14'30.60" would be described in DMS as: 43.040833° N and 77.241833° W.

Decimal	Deg Min Sec	Detail that can be identified
1.0	1° 00′ 0″	country or large region
0.1	0° 06′ 0″	large city or district
0.01	0° 00′ 36″	town or village
0.001	0° 00′ 3.6″	neighbourhood, street
0.0001	0° 00′ 0.36″	individual street, land parcel
0.00001	0° 00′ 0.036″	individual trees
0.000001	0° 00′ 0.0036″	individual humans
0.0000001	0° 00′ 0.00036″	limit of commercial surveying
0.00000001	0° 00′ 0.000036″	specialized surveying

Map Projections

These are systems which are used in mapping to show the curved surface of the Earth on a flat map; these are some of the most common:

UTM Universal Transverse Mercator (worldwide)
MGRS Military grid reference System (NATO)
USNG United States National Grid (United States)
OSNG Ordnance Survey National Grid (UK)
GARS Global Area Reference System (US Dept of Defense)
GEOREF World Geographic reference System (air navigation)
LCC Lambert Conformal Conic Projection (used for great circles in flying)
WGS84 World Geodetic System (used in GPS)
ECEF and ECR Earth Centered, Earth Fixed and Earth Centered Rotational

Most use eastings and northings relative to an arbitrary reference point at the bottom left hand corner of the map but WGS and ECEF use X, Y, Z coordinates related to a datum at the centre of the mass of the Earth.

THE SEVEN NATURAL FACTORS OF GREEK MEDICINE

The four elements

- Fire - igneous
- Air - gaseous
- Earth - solid
- Water – liquid

The four humors

- Blood – air
- Phlegm – water
- Yellow bile – fire
- Black bile – earth

The four temperaments

- Sanguine – blood
- Choleric – yellow bile
- Melancholic – black bile
- Phlegmatic – phlegm

The four faculties

- Vital – heart
- Natural – liver
- Psychic – brain
- Generic – gonads

The vital principles

- The vital force – kinetic energy
- The innate force – thermal energy
- Thymos - the immune force

The organs and parts of the body

Heart, lungs, throat, liver and gall bladder, stomach, spleen, intestines and bowel, colon, kidneys, adrenal glands, reproductive organs and the brain.

The administering virtues

- Sanguine – attractive virtue
- Choleric – digestive virtue
- Melancholic – retentive virtue
- Phlegmatic – expulsive virtue

The four basic qualities

Hot, cold, dry and wet

HERALDS

The Three English Kings at Arms

    Garter (Blue)   Principle in England
    Clarenceaux   Southern England
    Norroy and Ulster (Purple)   N England and N Ireland

The Six English Heralds

- Somerset
- Richmond
- Lancaster
- Windsor
- Chester
- York

The Four Pursuivants

- Rouge Dragon
- Blue Mantle
- Portcullis
- Rouge Croix

The Scottish King at Arms

- Lord Lyon

The Three Scottish Heralds

- Albany
- Marchmont
- Rothesay

The Five Scottish Pursuivants
- Unicorn
- Carrick
- Dingwell (Kintyre)
- Linlithgow
- Falkland

INDEXES

- KLCI   Kuala Lumpur
- **CAC 40**   **Paris**
- OMX   Stockholm
- FTSE/JSE   Johannesburg
- SMI and SPI   Zurich
- **FTSE 100**   **London**
- FTSE All Share   London
- All Ordinaries   Sydney
- BSE Sensex   Mumbai
- BEL 20   Brussels
- IBrX   Sao Paulo
- JCI or IHSG   Djarkarta
- **MIB**   **Milan**
- SET   Bangkok
- PSEI   Manilla
- ISEQ   Dublin
- WIG   Warsaw
- PSI   Lisbon
- **DAX**   **Frankfurt**
- NSX50   Auckland
- **Straights Times**   **Singapore**
- SET   Bangkok
- **Hang Seng**   **Hong Kong**
- S&P/TSX   Toronto
- **Dow Jones**   **New York**

- **S & P 500**   New York
- **NASDAQ**   New York
- OBX   Oslo
- TA – 125   Tel Aviv
- SSE   Shanghai
- **Nikkei 225**   **Tokyo**
- MICEX   Moscow
- Bolsa   Mexico City
- **IBEX**   **Madrid**

There are now many share indexes; the above are just a few of the main ones. The best known ones are shown in bold.

MUSIC AND SOUND

Musical groups

- Soloist   1
- Duet   2
- Trio   3
- Quartet   4
- Quintet   5
- Sextet   6
- Septet   7
- Octet   8

Sound levels (the descriptions are not definitive)

Each rise of 10 decibels increases the sound level tenfold

0	inaudible
20	soft whisper
30	whisper
40	quiet speech
50	quiet at home, normal conversation
60	light traffic, close conversation
70	busy street
80	light machinery
90	heavy traffic, road drill, heavy machinery
100	plane taking off at half a km
115	loud disco

120   near to plane taking off
130   pain threshold
140   alongside a plane taking off

PAPER AND BOOK SIZES

Standard paper sizes

A0	841 x 1189 mm
A1	594 x 841
A2	420 x 594
A3	297 x 420
A4	210 x 297
A5	148 x 210
A6	105 x 148
A7	74 x 105
A8	52 x 74
A9	37 x 52
A10	26 x 37
B0	1000 x 1414 mm
B1	707 x 1000
B2	500 x 707
B3	353 x 500
B4	250 x 353
B5	176 x 250
B6	125 x 176
B7	88 x 125
B8	62 x 88
B9	44 x 62

B10	31 x 44	
C0	917 x 1297 mm	
C1	648 x 917	
C2	458 x 648	
C3	324 x 458	
C4	229 x 324	
C5	162 x 229	
C6	114 x 162	
C7	81 x 114	
DL	110 x 220	
C7/6	81 x 162	

Imperial paper sizes

Imperial	559 x 762 mm	22 x 30 in
Elephant	508 x 686	20 x 27
Royal	508 x 635	20 x 25
Medium	457 x 584	18 x 23
Demy	445 x 572	17.5 x 22.5
Large post	419 x 533	16.5 x 21
Crown	381 x 508	15 x 20
Foolscap	343 x 432	13.5 x 17

Book sizes

Foolscap octavo	170 x 110 mm	6.75 x 4.25 in
Crown octavo	190 x 125	7.50 x 5.00
Large crown octavo	205 x 135	8.00 x 5.25
Small demy octavo	215 x 145	8.50 x 5.63
Demy octavo	220 x 145	8.75 x 5.63

Medium octavo	230 x 145	9.00 x 5.75
Small royal octavo	235 x 155	9.25 x 6.13
Royal octavo	255 x 160	10.0 x 6.25

The commonest in current use are 5"x 8" and 6"x 9".

RHYMES AND SONGS

One, two......

    One, two, buckle my shoe

    Three four, knock at the door (shut the door)

    Five, six, pick up sticks

    Seven, eight, lay them straight

    Nine, ten, a big fat hen (a good fat hen)

    Eleven. twelve, dig and delve (who will delve?)

    Thirteen, fourteen, maids a-courting

    Fifteen, sixteen maids in the kitchen (maids a kissing)

    Seventeen, eighteen, maids a waiting

    Nineteen, twenty, my plate's empty (stomach's empty)

Numbers (see also 'Magpies')

    One for Sorrow

    Two for Joy

    Three for a Girl

    Four for a Boy

    Five for Silver

    Six for Gold

    Seven is a Secret never to be told

    Eight's a Wish

    Nine is a Kiss

    Ten is a Bird you must not miss

Magpies

    One's sorrow

    Two's mirth

    Three's a wedding

    Four's a birth

Five's a christening (for silver)
Six a death (for gold)
Seven's heaven (for a secret, not to be told)
Eight is hell (for heaven)
Nine is for hell and
Ten for the Devil's own sel

Fish

One, two, three, four, five
Once I caught a fish alive
Why did you let it go?
Because it bit my finger so
Six, seven, eight, nine, ten
Then I let it go again
Which finger did it bite
This little finger on the right

Threesomes

Three young rats, with black felt hats
Three young ducks with white straw flats
Three young dogs with curling tails
Three young cats with demi-veils
Went out to walk with two young pigs
In satin vests and sorrel wigs
But suddenly it chanced to rain
And so they all went home again

A Numbers Ditty

One's none; Two's some
Three's a many; Four's plenty
Five's a little hundred

Peter and Paul

> Two little dicky birds, sitting on a wall
> One named Peter, one named Paul
> Fly away Peter, fly away Paul
> Come back Peter, come back Paul

White Horses

> Thirty white horses, upon a red hill
> Now they tramp, now they champ
> Now they stand still

Like a Teddy Bear

> Round and round the garden, like a Teddy bear
> One step, two step, tickle him under there

The Wide-mouthed Waddling Frog

> Twelve huntsmen with horn and hounds, hunting over other mens' grounds
> Eleven ships sailing o'er the Main, some bound for France and some for Spain
> I wish them all safe home again
> Ten comets in the sky, some low and some high
> Nine peacocks in the air; I wonder how they all came there
> I don't know and I don't care
> Eight joiners in a joiners hall, working with their tools and all
> Seven lobsters on a dish as fresh as any heart could wish
> Six beetles against a wall, close by an old woman's apple stall
> Five puppies by our dog Ball, who daily for their breakfast call

Four horses, stuck in a bog
Three monkeys tied to a clog
Two pudding ends would choke a dog
With a gaping, wide-mouthed waddling frog

My Black Hen

Hickety, pickety, my black hen
She lays eggs for gentlemen
Gentlemen come every day
To see what my black hen doth lay
Sometimes nine and sometimes ten
Hickety, pickety, my black hen

Kittens and Mittens

Three little kittens had lost their mittens…………………..

Mice

Six little mice sat down to spin……………………

Three blind mice, see how they run…………………….

Pigeons

I had two pigeons bright and gay
They flew from the other day
………………………

Hens

There was an old man who lived in Middle Row
He had five hens and a name for them, oh!
Bill and Ned and Battock

Cut-her-foot and Pattock
Chuck, my lady Pattock
Go to thy nest and lay

Bah Bah Black Sheep

Bah bah black sheep, have you any wool?
Yes Sir, yes Sir, three bags full
One for the Master and one for the Dame
And one for the little boy who lives down the lane

A Cow

Four stiff standers, four dilly danders
Two lookers, two crookers, and a wig wag

Billy loves tea

One, two, three
I love coffee and Billy loves tea
How good you be
One, two, three
I love coffee and Billy loves tea

Mary

One, two, three, four
Mary at the cottage door
Five, six, seven, eight
Eating cherries off a plate

Jesus

One, two, three, four, five
Jesus Christ is still alive
Six, seven, eight, nine, ten
He will come back again

Miss One, Two, Three

>Miss One, Two, Three could never agree
>While they gossiped around a tea caddy

The Cooks of Colebrook

>There were three Cooks of Colebrook
>And they fell out with our cook
>And all was for a pudding he took
>From the three Cooks of Colebrook

Gregory Griggs

>Gregory Griggs, Gregory Griggs
>Had forty seven different wigs
>He wore them up, he wore them down
>To please the people of Boston town
>He wore them East, he wore them West
>But he never could tell which he loved best

Angels

>Matthew, Mark, Luke and John
>Bless the bed that I lie on
>Four corners to my bed
>Four angels round my head
>One to watch and one to pray
>And two to bear my soul away

Bedtime

>Go to bed first – a golden purse
>Go to bed second – a golden pheasant
>Go to bed third – a golden bird

Three Men in a Tub

> Rub-a-dub-dub, three men in a tub
> And how do you think they got there?
> The butcher, the baker, the candlestick-maker
> They all jumped out of a rotten potoato
> 'Twas enough to make a man stare

Duke of York

> Oh, the brave old Duke of York, he had ten thousand men
> He marched them up to the top of the hill, and he marched them down again

The King of France

> The King of France, the King of France, with forty thousand men
> Oh, they all went up the hill, and so – they all came down again

Three Sons

> There was a woman had three sons
> Jerry and James and John
> Jerry was hung and James was drowned
> John was lost and never found
> So there was an end of her three sons
> Jerry and James and John

Ten Little Indians (one version)

> One goes home and then there were nine
> One falls off a gate and then there were eight
> One goes to sleep and then there were seven
> One breaks his neck and then there were six

One kicks the bucket and then there were five
One falls into the cellar and then there were four
One gets fuddled and then there were three
One falls out of a canoe and then there were two
One gets shot by the first and then there was one
One gets married and then there were none

St Ives

As I was going to St Ives, I met a man with seven wives
Each wife had seven sacks, each sack had seven cats
Each cat had seven kits
Kits, cats, sacks and wives, how many were going to St Ives?

Correctly 1 but the number coming from St Ives is also interesting:

Man	*1*
Wives	*7*
Sacks	*49*
Cats	*343*
Kits	*2401*
Total	*2801*

LNB's

Ten lnb's went out to dine, one choked his little self and then there were nine

Nine lnb's sat up very late, one overslept himself and then there were eight

Eight lnb's travelling in Devon, one said he'd stay there and then there were seven

Seven lnb's chopping up sticks, one chopped himself in half and then there were six

Six lnb's playing with a line, a bumble bee stung one and then there were five

Five lnb's going in for law, one got in chancery and then there wer four

Four lnb's going out to sea, a red herring swallowed one and then there were three

Three lnb's walking in the zoo, a big bear hugged one and then there were two

Two lnb's sitting in the sun, one got frizzled up and then there was one

One lnb living all alone, he got married and then there were none

Sailors

Four and twenty sailors went to kill a snail
………………..

Babylon

How many miles to Babylon?
Three score miles and ten………..

Cripplegate

Three crooked cripples went through Cripplegate
And through Cripplegate went three crooked cripples

Banbury Cross

Ride a cock horse to Banbury Cross
To see what Tommy can buy
A white penny loaf, a white penny cake
And a two-penny apple pie

April

> March borrowed frae Aprile
> Three days an' they were ill
> The first o' them was wind and weet
> The second o' them was snow an' sleet
> The third o' them was sic a freeze
> That the birds legs stack to the trees

The Twelve Days of Christmas

> A partridge in a pear tree
> Two turtle doves
> Three French hens
> Four <u>colly</u> birds
> Five gold rings
> Six geese a-laying
> Seven swans a-swimming
> Eight maids a-milking
> Nine drummers drumming
> Ten pipers piping
> Eleven ladies dancing
> Twelve lords a-leaping

St Swithin

> St Swithin's Day, if thou doth rain, for forty days it will remain.
> St Swithin's Day, if thou be fair, for forty days 'twill rain no more

Days in the Months

> Thirty days hath September, April, June and November
> All the rest have thirty one, excepting February alone

And that has 28 days clear and twenty nine in each leap year

Thirty days hath September, April, June and November
All the rest have thirty one, excepting leap year
Thats the time, when February's days are twenty nine

But thirty days November hath, April, June and September
February hath but twenty eight without a leap attender

Work

He that would thrive must rise at five
He that hath thriven may lie in till seven
And he that by the plough would thrive
Himself must either hold or drive

Five Outs

Out of money, and out of clothes
Out of heels, and out at the toes
Out of credit – but don't forget,
Never out of but aye in debt!

Riding

One to make ready
And two to prepare
Good luck to the rider
And away with the mare

'ery nonsense

One-ery, two-ery, tickery, seven

Hallibo, crackibo, ten and eleven
Spin, span, muskidan
Twiddle-um, twaddle-um, twenty one

Green Grow the Rushes-oh
1   All alone and evermore shall be so
2   Lilywhite boys
3   Rivals
4   Gospel makers
5   Symbols at your door
6   Proud walkers
7   Stars in the sky
8   Bold rangers
9   Bright shiners
10 Commandments
11 Who went to Heaven
12 Apostles

Hot Cross Buns

One a penny, two a penny, hot cross buns
If your daughters do not like them, give them to your sons
But if you haven't any of those pretty little elves
You cannot do better than eat them yourselves

Teeth

Four and twenty white bulls
Sat upon a stall
Forth came the red bull
And licked them all

Mispunctuation

        Every lady in this land
        Has twenty nails / upon each hand
        Five / and twenty on hands and feet
        All this is true without deceit

Riddle

        A riddle, a riddle, as I suppose
        A hundred eyes and never a nose

Love

        One I love, two I love, three I love I say
        Four I love with all my heart, five I cast away
        Six he loves, seven she loves, eight both love
        Nine he comes, ten he harries
        Eleven he courts, twelve he marries

Legs

        Two legs sat upon three legs
        With one leg in his lap
        In comes four legs
        And runs away with one leg
        Up jumps two legs
        Catches up three legs
        Throws it after four legs
        And makes him bring back one leg
        (man = 2, mutton = 1, stool = 3, dog = 4)

Conundrum

        Twelve pears hanging high
        Twelve knights riding by
        Each knight took a pear
        And yet left eleven there

Ship

    I saw a ship a sailing, a-sailing on the sea
    And oh but it was laden with pretty things for thee
    There were comfits in the cabin and apples in the hold
    The sails were made of silk and the masts were all of gold
    The four-and-twenty sailors that stood between the decks
    Were four-and-twenty white mice with chains about their necks
    The captain was a duck with a packet on his back
    And when the ship began to move the captain said quack! Quack!

Sleep

    Nature requires five, custom gives seven
    Laziness takes nine and wickedness eleven

Peas pudding

    Pease-pudding hot, pease-pudding cold
    Pease-pudding in the pot, nine days old
    …………………

Sixpence

    Sing a song of sixpence, a pocket full of rye
    Four and twenty blackbirds, baked in a pie

    I love sixpence, jolly little sixpence……………………

The Moon

    There was a thing a full month old
    When Adam was no more
    Before the thing was five weeks old
    Adam was years four score

Proverbs

Jack of all trades and master of **none**

One fool makes many

One good turn deserves another

One man's meat is another man's poison

One swallow does not make a summer

Don't put all of your eggs in **one** basket

Every dog is allowed **one** bite

In the land of the blind, the **one** eyed man is king

Of **one** ill come many

There's more than **one** way to skin a cat

One today is worth **two** tomorrows

If a man deceive me **once**, shame on him; if **twice**, shame on me

To kill **two** birds with **one** stone

To take **two** bites at the cherry

Two wrongs do not make a right

Two eyes see better than **one**

Two heads are better than **one**

There are **two** sides to every question

A bird in the hand is worth **two** in the bush

It takes **two** to make a quarrel

Masters **two** will not do

Of **two** evils choose the less

He giveth **twice** that gives in a trice

Opportunity seldom knocks **twice**

One cannot die **twice**

He who pays last never pays **twice**

Two is company, **three** is none (**three**'s a crowd)

From clogs to clogs in only **three** generations

Six of one and half a dozen of the other

Rain before **seven**; fine before **eleven**

A stitch in time saves **nine**

A cat has **nine** lives

A wonder lasts **nine** days

Nine tailors make but one man

One may lead a horse to water but **twenty** cannot make him drink

There's only **twenty four** hours in the day

A fool at **forty** is a fool indeed

Life begins at **forty**

It will all be the same in a **hundred** years

A pullet in the pen is worth a **hundred** in the fen

There's safety in numbers!

ROMAN NUMERALS

The basic numbers

1	I
5	V
10	X
50	L
100	C
500	D
1000	M

(Overlining makes the letter one thousand bigger)

5,000	V̄
10,000	X̄
100,000	C̄

Sample Dates:

1000 = M   1066 = MLXVI
1100 = MC   1200 = MCC
1300 = MCCC   1400 = MCD
1500 = MD   1600 = MDC
1666 = MDCLXVI (Great Fire of London)
1700 = MDCC   1800 = MDCCC
1900 = MCM   1990 = MCMXC
1991 = MCMXCI   1992 = MCMXCII
1993 = MCMXCIII   1994 = MCMXCIV
1995 = MCMXCV   1996 = MCMXCVI
1997 = MCMXCVII   1998 = MCMXCVIII
1999 = MCMXCIX   2000 = MM

SCALES AND GRADING SYSTEMS

The Beaufort Wind scale

No.	Land	kph	Description
0	Calm	<1	smoke rises vertically
1	Light air	1-5	smoke drifts
2	Light breeze	6-11	felt on face
3	Gentle breeze	12-19	extends light flag
4	Moderate breeze	20-28	raises dust and light paper
5	Fresh breeze	29-38	small trees sway
6	Strong breeze	39-49	umbrellas difficult
7	Near/moderate gale*	50-61	walking inconvenient
8	Gale/fresh gale*	62-74	twigs break from trees
9	Strong gale	75-88	chimney pots, slates fall
10	Storm/whole gale*	89-102	trees uprooted
11	Violent storm/storm*	103-114	widespread damage
12 to 17	Hurricane	117+	extremely violent

* descriptions vary according to source

Modified Mercalli Earthquake Scale

Mod Mercalli	Approx Richter	Observation
I	>3.0	Not felt
II	3.5	Felt on upper floors
III	4.2	Felt indoors, objects swing
IV	4.5	Cars rock, trees shake,
V	4.8	Doors swing, liquids spill
VI	5.4	Windows break, pictures fall
VII	6.1	Standing difficult, plaster falls
VIII	6.5	Masonry damage, tree branches broken
IX	6.9	Masonry destroyed, general panic
X	7.3	Most buildings destroyed, landslides
XI	8.1	Rails bend, roads /pipelines broken up
XII	8.2+	Total destruction, rock masses displaced

Mohs Hardness Scale

1	Talc	crushed with finger nail
2	Gypsum	scratched with finger nail
3	Calcite	scratched with copper coin
4	Fluospar	scratched by glass
5	Apatite	scratched by penknife
6	Feldspar	scratched by quartz
7	Quartz	scratched by steel file
8	Topaz	scratched by corundum
9	Corundum	scratched by diamond
10	Diamond	

Richter Earthquake Scale

This is a theoretical (logarithmic) scale with no common descriptions or limits.

0 is at rest and 10 is viewed as a practical maximum.

Each step is 10 times stronger than the one below, e.g 6.0 is 10 times stronger than 5.0

Kinsey Scale of Homsexuality

- 0 Exclusively hetero
- 1 Predominantly hetero, only incidentally homosexual
- 2 Predominantly hetero, more than incidentally homosexual
- 3 Equally hetero and homosexual
- 4 Predominantly homosexual, more than incidentally hetero
- 5 Predominantly homosexual, only incidentally hetero
- 6 Exclusively homosexual
- X No socio-sexual contacts or reactions

The Wave Scale

Number	description	height (ft)
0	calm	0.25
1	calm	0.5 - 1
2	smooth	2-3
3	smooth	3-5
4	slight	6-8
5	moderate	9-13
6	rough	13-19

7	very rough	18-25
8	high	23-32
9	very high	29-41
10	very high	37-52
11	phenomenal	45+
12	phenomenal	++

The Douglas Wave and Swell Scale (International)

No.	Description	Wave Index
0	calm	00
1	smooth	10
2	slight	20
3	moderate	30
4	rough	40
5	very rough	50
6	high	60
7	very high	70
8	precipitous	80
9	confused	90

Swell figures of 0-9 are added to index to give figures in the range 00 to 99

Alcohol content (for spirits)

UK Proof	US proof	% alcohol
175.0	200.0	100.0
100.0	114.3	57.1
87.5	100.0	50.0
85.0	97.1	48.6

80.0	91.4	45.7
75.0	85.7	42.9
70.0	80.0	40.0
65.0	74.3	37.1
60.0	68.6	34.3
52.5	60.0	30.0
50.0	57.1	28.6
35.0	40.0	20.0
26.3	30.0	15.0
21.0	24.0	12.0
17.4	20.0	10.0

Temperature

There are three scales though only two of them are in use today, Celsius (which used to be known as 'centigrade') and Fahrenheit. The third (Reaumur) had a 0 to 80 scale from freezing to boiling water.

Degrees C	**Degrees F**	**Description**
-278		Absolute zero
-40	-40	The only point where they are the same
0		Approx freezing point of salt water
0	32	Water Freezes
28	82	Numeric transposition
37	98.4	Body temperature
100	212	Boiling water
	451	Spontaneous ignition of paper

To convert from F to C: deduct 32 then multiply by 5 then divide by 9.
To convert from C to F: multiply by 9 then divide by 5 then add 32

Credit Rating Scores

850	Highest possible
750+	Excellent
700-749	Good
650-699	Fair
600-649	Poor
< 600	Bad
300	Lowest possible

Wine Ratings

Parker		Robinson	5 star
95-100	Classic	19-20 Truly exceptional	5*
90-94	Outstanding	18 A humdinger and 17 A cut above superior	5* 4*
85-89	Very good	16 Superior	3*
80-84	Good	15 Average	2*
75-79	Mediocre	14 Deadly dull	1*
50-74	Not recommended		1*

The Scoville Heat Scale

This scale, which is named after Wilbur L Scoville, is based on the dilution factor necessary for the majority of an experienced tasting panel to discern no heat and is measured in Scoville Heat Units (SHUs). It can be refined to cover aspects of how heat is experienced such as: development over time; duration; location (lips, mouth, throat, stomach); sensation of heat and intensity. It varies from 100 or less (e.g. sweet peppers) to over 3 million for the very hottest and is highly subjective. It would possibly be better described in a log scale.

SHUs	Typical peppers/chillis
0-100	Bell peppers, Thai peppers, Cayenne
100 - 1000	Peperonicini, Pimento
1,000 – 2,000	Anaheim, Poblano, Rocotillo
2,000 – 5,000	Jalapeno, Tabasco sauce
5,000 – 15,000	Wax Peppers, Serrano, Guajillo, Mirasol,
15k – 30k	Hidalgo, Hot Wax peppers,
30k – 50k	Cayenne, Aji
50k – 100k	Thai peppers, Malagueta
100k – 300k	Fatali, Poblano, Scotch Bonnet, Habanero, Jamaican Hot
300k – 900k	Habanero varieties, Fatali, Birds Eye, Jalapeno, Dorset Naga
900k – 2m	Ghost Pepper, Trinidad Scorpion, Carolina Reaper, Dragon's Breath
2m – 5m	Pepper spray

The Holmes and Rahe Stress Scale

	Weight	Life Change Unit (LCU)
1	100	Death of spouse
2	73	Divorce
3	65	Marital separation
4	63	Jail term
5	63	Death of close family member
6	53	Personal injury or illness
7	50	Marriage
8	47	Fired at work
9	45	Marital reconciliation
10	45	Retirement
11	44	Change in health of family member
12	40	Pregnancy
13	39	Sex difficulties
14	39	Gain of new family member
15	39	Business readjustment
16	38	Change in financial state
17	37	Death of close friend
18	36	Change to a different line of work
19	35	More arguments with spouse
20	35	A large mortgage or loan
21	31	Foreclosure of mortgage or loan
22	30	Change in responsibilities at work
23	29	Son or daughter leaving home
24	29	Trouble with in-laws
25	28	Outstanding personal achievement
26	26	Spouse begins or stops work
27	26	Begin or end school/college
28	25	Change in living conditions

29	24	Revision of personal habits
30	23	Trouble with boss
31	20	Change in work hours or conditions
32	20	Change in residence
33	20	Change in school/college
34	19	Change in recreation
35	19	Change in church activities
36	18	Change in social activities
37	17	A moderate loan or mortgage
38	16	Change in sleeping habits
39	15	Change in number of family get-togethers
40	15	Change in eating habits
41	13	Vacation
42	12	Christmas
43	11	Minor violations of the law

The score (weighting) for each LCU which has been experienced is added to make a total which is then used to indicate the likelihood of illness occurring in the individual.

Score	Risk of becoming ill in the near future
300-600	High or very high risk
150-299	Moderate risk
11-149	Low risk

Moral Behaviour

The grades for moral behaviour are loosely based on those which can be found in a number of religious books though they tend not to use numerical values. They epitomize the grading systems which work well as opposed to those that don't.

1	Rewardable	Duty
2	Laudable	Merit
3	Acceptable	Permissible
4	Reprehensible	Reprehensible
5	Punishable	Forbidden

Maturity

1	Innocent	Starting to learn
2	Aware	Starting to apply
3	Developing	Developing and embedding
4	Competent	Developed and embedded…
5	Effective	…and incorporated
6	Excellent	Integrated and continuously improved

Prosperity, availability and attainment

These are cited as examples of how a grading system may be developed to cover any eventuality.

Grade	Prosperity	Availability	Attainment
1	Fortune	Abundant	Miracle
2	Prosperity	Plenty	Victory
3	Comfort	Available	Draw
4	Poverty	Scarce	Disaster
5	Ruination	Absent	Catastrophe

There is, of course for pedants, the necessity to have all of the words in the same case – nouns, verbs, adverbs or adjectives.

Food Hygiene Rating

　　0　Urgent improvement necessary
　　1　Major improvement necessary
　　2　Improvement necessary
　　3　Generally satisfactory
　　4　Good
　　5　Very good

This is a poor system as it confuses actions and grades; also the '0' is unnecessary.

State of alert against terrorist threats in the UK

This is an indicator of the state of readiness of the security forces based on intelligence relating to terrorist threats.

　　1　Low　　　　　　An attack is unlikely
　　2　Moderate　　　An attack is possible but not likely
　　3　Substantial　　An attack is a strong possibility
　　4　Severe　　　　An attack is highly likely
　　5　Critical　　　　An attack is expected imminently

The level is set by the Joint Terrorism Analysis Centre and the Security Service (MI5). They use only the descriptive word, not numbers, codes or colours.

DEFCON (Defence Condition)

These are used to indicate the state of readiness of the American armed forces.

Readiness	Code name	Description	Readiness	Color
1	Cocked pistol	War imminent	Maximum	White
2	Fast pace	Above normal	Ready to deploy	Red
3	Round house	Increase in force	Ready to mobilize	Yellow
4	Double take	Watch and strengthen	Above normal	Green
5	Fade out	Lowest state	Normal	Blue

Films

The British Board of Film Classifications is based on the age that is appropriate for film goers. In its simplest form:

U	Universal	All ages
PG	Parental guidance	All ages but some scenes not suitable for younger children
12A	12s and over	Not suitable for under 12s but may be OK if accompanied by an adult
12	12 and over	Only suitable for age 12 and over
15	15 and over	Only suitable for age 15 and over
18	18 and over	Only suitable for age 18 and over
R18	18 and over	Only suitable for age 18 and over in adult cinemas

Asset Management

'Asset Management' is the methodology used to determine when physical assets need to be refurbished or replaced as they near the end of their working life in order that they may continue to provide outputs for the business which owns them. The five grades are based on those developed by the UK's Water Research Centre for the classification of pipelines which was first introduced in the 1970s.

1 or A	Excellent	Fit for the future
2 or B	Good	Adequate for now
3 or C	Adequate	Requires attention
4 or D	Poor	At risk
5 or E	Awful	Unfit for purpose

The grades are used for both condition and performance (based on levels of service).

Air Quality Index (UK)

There are many different systems in use according to the country to which they apply. This is a simplification of that which applies in the UK:

Index	NOX	PM(10)	Risk	Action
1	< 67	< 16	Low)	
2	< 134	< 30	Low)	Enjoy usual activity
3	< 200	< 50	Low)	
4	< 267	< 58	Moderate]	
5	< 334	< 66	Moderate]	Enjoy usual activity
6	< 400	< 75	Moderate]	

7	< 467	< 83	High)	Reduce outdoor activity if suffering discomfort
8	< 534	< 91	High)	
9	< 600	< 100	High)	
10	>= 601	>= 101	Very High	Reduce outdoor activity

Intelligence

There are about a dozen ranking systems based on IQ which can be found on line. The simplified example below is based on Terman's Stanford-Binet 1916 classification.

IQ range	Classification
Above 140	Genius or near genius
120 - 140	Very superior
110 - 120	Superior
90 – 110	Normal or average
80 – 90	Dull, rarely classifiable as feeble minded
70 – 80	Sometimes dull often feeble minded
Below 70	Feeble minded

British University degrees

First class honours (with distinction)
Upper second class honours
Lower second class honours
Third class honours
Ordinary (Pass degree)

American University Degrees

Grade	Scale (%)	UK equivalent
A+	90-100	Distinction
A	80+	Distinction
A-	70+	
B+	60+	First class
B	50+	Second class
B-	45+	
C+	40+	Third class
C	32+	Third class
C-		
D+		
D		
F	<32	Fail

Qualification

Level	Typical designation	Education	Remarks
7	Consultant/technical specialist	PhD plus specialization	Equivalent to professor
6	Executive director	Higher degree/MBA	Equivalent to doctorate
5	Senior manager	Degree	Graduate
4	Middle manager	Diploma/BTEC	Technical diploma
3	Supervisor	NVQ3	
2/3	Craftsman	City and Guilds	Plumbers, electricians, carpenters etc.

| 2 | Senior operative | NVQ2 |
| 1 | Operative/manual worker | NVQ1 |

These are typical for the UK and there is, obviously, a high degree of overlap between the levels. A slightly different version incorporating eight grades may be found on Wikipedia.

OFSTED Ratings for Schools

Arguably this system does not work well as it has no grade for a school which is performing 'adequately' (which would presumably make up the majority).

1 = outstanding
2 = good
3 = requires improvement
4 = inadequate (in special measures)

GCSE

Old system	new
A*	9
A	8 and 7
B	6 and 5
C	5 and 4
D	3
E	3
F	2
G	1
U	1

The Bristol Scale for Faeces

There are pictures to illustrate these but we won't confront you with them. The seven types of stool are:

- Type 1: Separate hard lumps, like nuts (hard to pass)
- Type 2: Sausage-shaped, but lumpy
- Type 3: Like a sausage but with cracks on its surface
- Type 4: Like a sausage or snake, smooth and soft
- Type 5: Soft blobs with clear cut edges (passed easily)
- Type 6: Fluffy pieces with ragged edges, a mushy stool
- Type 7: Watery, no solid pieces, entirely liquid

COGIATI (Combined Gender Identity and Trans Sexuality Inventory)

Classification	Description
1	Standard male
2	Feminine male
3	Androgyne
4	Probable transsexual
5	Transsexual

Burns

The British system measures burns according the severity (depth) of the injury:

First degree	Superficial	Epidermal
Second degree		Deep dermal
Third degree		Subcutaneous
Fourth degree	Underlying	Affecting fat muscle and bone

The American system of burn severity is based on the area of coverage:

	Adult	**Young and old**
Minor	<10% TBSA	<5% TBSA*
	<2% full thickness burn	
Moderate	10-20% TBSA	
	2-5% full thickness burn	
Major	>20% TBSA	
	>5% full thickness burn	

*Total Body Surface Area

Possible side effects (from drugs)

Common	may affect up to 1 in 10 people
Uncommon	may affect up to 1 in 100 people
Rare	may affect up to 1 in 1,000 people
Very rare	may affect up to 1 in 10,000 people

Walking and climbing

A simple grading system for walks is based on just three grades:

1 – Easy – easy stroll on roads and paths, level walk, no hills

2 – Moderate – fairly easy walk on roads and paths, some inclines

3 – Hard – fairly strenuous, several inclines/hills on roads and paths

This might be OK for some basic walks but would hardly suffice in any National Park.

The more complex system now divides all hikes and climbs into five classes: The exact definition of the classes is somewhat controversial, and updated versions of these classifications have been proposed.

Class 1: Walking with a low chance of injury, hiking boots a good idea.

Class 2: Simple scrambling, with the possibility of occasional use of the hands. Little potential danger is encountered. Hiking Boots highly recommended.

Class 3: Scrambling with increased exposure. Handholds are necessary. A rope should be available for learning climbers, and if you just choose to use one that day, but is usually not required. Falls could easily be fatal.

Class 4: Simple climbing, with exposure. A rope is often used. Natural protection can be easily found. Falls may well be fatal.

Class 5: Is considered technical roped free (without hanging on the rope, pulling on, or stepping on anchors) climbing; belaying, and other protection hardware is used for safety. Un-roped falls can result in severe injury or death.

[Class 5.0 to 5.12 + is used to define progressively more difficult free moves.]

Class 6: Is considered Aid (often broken into A.0 to A.5) climbing. Equipment (Etriers, aiders, or stirrups are often used to stand in, and the equipment is used for hand holds) is used for more than just safety.

European Hotelstars Union

One star	Tourist
Two star	Standard
Three star	Comfort
Four star	First class
Five star	Luxury

The stars are awarded on the facilities available rather than on the quality of that service. An 'S' after the stars indicates a 'superior' level within the basic grade.

UNITS AND PREFIXES

Basic Units

m	meter	length
kg	kilogram	mass
s	second	time
A	ampere	current
K	kelvin	temperature
mol	mole	amount
cd	candela	luminosity

Secondary (derived) Units

Hz	Hertz	frequency
N	Newton	force
Pa	Pascal	pressure or stress
J	Joule	energy and heat
W	Watt	power, radiant flux
C	Coulomb	electrical quantity or charge
V	Volt	electrical potential
F	Farad	capacitance
W	Ohm	electrical resistance
S	Siemens	conductance
Wb	Weber	magnetic flux
T	Tesla	magnetic flux density
H	Henry	inductance
lm	Lumen	luminous flux
lx	lux	illuminance

Supplementary Units

rad	radian	angle
sr	steradian	solid angle

Biblical Units

The units found in the Bible depend, for their names and conversions, as much on the translators as on the text itself and should not, therefore, be relied upon.

Length

Finger	18mm		Jeremiah 52:21
Hand	= 4 fingers	7-8 cm	Exodus 25:21
Span	23cm		Exodus 28:16
Cubit	= 2 spans	46cm	numerous
Fathom	1.8m		Acts 27:28
Reed	= 6 cubits	2.7m	Ezekiel 40:5
Furlong	200m		Revelation 14:20

Weight (especially of silver)

Gerah	0.6gr		Ezekiel 45:12
Bekah	= 10 gerahs	6gr	Exodus 38:26
Pim	4gr		Samuel 13:21
Shekel	= 2 bekah	12gr	Genesis 24:22, Exodus 38:26
Shekel	= 3 pim	12gr	Samuel 13:21
Shekel	= 20 gerahs	12gr	Exodus 30:13, Numbers 3:47
Maneh	= 50 shekels	600gr	Ezra 2:69
Talent	= 60 manehs	34kg	numerous
Talent	= 3,000 shekels		

Volume (wet)

Log	0.3 lit		Leviticus 14:10
Kab or cab	= 4 logs	1.2 lit	2 Kings 6:25
Hin	= 12 logs	3.7 lit	Numbers 15:4
Bath	= 6 hins	22 lit	Isaiah 5:10
Ephah	= 6 hins	22 lit	Ezekiel 45:11
Homer	= 10 baths	220 lit	Ezekiel 45:11
Kor or cor	= 10 baths	220 lit	1 Kings 5:11

Volume (dry)

Omer	2.2 lit		Exodus 16:36
Ephah	= 10 omers	22 lit	Ruth 2:17
Homer	= 10 ephahs	220 lit	Leviticus 27:16

Magnitude Prefixes

T	tera	10^{12}
G	giga	10^9
M	mega	106
k	kilo	103
h	hecto	102
da	deca	101
d	deci	10-1
c	centi	10-2
m	milli	10-3
u	micro	10-6
n	nano	10-9
p	pico	10-12
f	femto	10-15
a	atto	10-18

Computer memory

> 1 bit = 1 or 0
> a nibble = 4 bits
> 1 byte = 8 bits
> 1 kilobyte (KB) = 1024 bytes
> 1 megabyte (MB) = 1024 kilobytes
> 1 gigabyte (GB) = 1024 megabytes
> 1 terabyte (TB) = 1024 gigabytes
> 1 petabyte (PB) = 1024 terabytes
> 1 exabyte (EB) = 1024 petabytes
> 1 zettabyte (ZAB) = 1024 exabytes
> 1 yottabyte (YB) = 1024 zettabytes

Hellabyte and Brontobyte have been proposed should larger numbers be required.

VEHICLE REGISTRATION - SUFFIXES AND PREFIXES IN THE UK

	XXX 999 A	**A 999 XXX**
A	1963	1983/84
B	1964	1984/85
C	1965	1985/86
D	1966	1986/87
E	1967 up to August	1987/88
F	1967/68	1988/89
G	1968/69	1989/90
H	1969/70	1990/91
I	Not used	Not used
J	1970/71	1991/92
K	1971/72	1992/93
L	1972/73	1993/94
M	1973/74	1994/95
N	1974/75	1995/96
O	Not used	Not used
P	1975/76	1996/97
Q	Special registration	Special registration
R	1976/77	1997/98
S	1977/78	1998/99
T	1978/79	
U	Not used	
V	1979/80	
W	1980/81	

X   1981/82
Y   1982/83
Z   Not used

The current system

		Postal areas	DVLA memory tag identifier
A	**Anglia**	Peterborough, Norwich and Ipswich	AA AB AC AD AE AF AG AH AJ AK AL AM AN AO AP AR AS AT AU AV AW AX AY
B	**Birmingham**	Birmingham	BA – BY
C	**Cymru**	Cardiff, Swansea and Bangor	CA CB CC CD CE CF CG CH CJ CK CL CM CN CO CP CR CS CT CU CV CW CX CY
D	**Deeside to Shrewsbury**	Chester and Shrewsbury	DA DB DC DD DE DF DG DH DJ DK DL DM DN DO DP DR DS DT DU DV DW DX DY
E	**Essex**	Chelmsford	EA – EY
F	**Forest & Fens**	Nottingham and Lincoln	FA FB FC FD FE FF FG FH FJ FK FL FM FN FP FR FS FT FV FW FX FY
G	**Garden of England**	Maidstone and Brighton	GA GB GC GD GE GF GG GH GJ GK GL GM GN GO GP GR GS GT GU GV GW GX GY

H	**Hampshire & Dorset**	Bournemouth and Portsmouth	HA HB HC HD HE HF HG HH HJ HK HL HM HN HO HP HR HS HT HU HV HW HX HY
K		Borehamwood and Northampton	KA KB KC KD KE KF KG KH KJ KK KL KM KN KO KP KR KS KT KU KV KW KX KY
L	**London**	Wimbledon, B'hamwood and Sidcup	LA LB LC LD LE LF LG LH LJ LK LL LM LN LO LP LR LS LT LU LV LW LX LY
M	**Manchester & Merseyside**	Manchester	MA – MY (MN + MAN for Isle of Man)
N	**North**	Newcastle and Stockton	NA NB NC ND NE NG NH NJ NK NL NM NN NO NP NR NS NT NU NV NW NX NY
O	**Oxford**	Oxford	OA – OY
P	**Preston**	Preston and Carlisle	PA PB PC PD PE PF PG PH PJ PK PL PM PN PO PP PR PS PT PU PV PW PX PY
R	**Reading**	Theale	RA – RY
S	**Scotland**	Glasgow, Edinburgh, Dundee, Aberdeen and Inverness	SA SB SC SD SE SF SG SH SJ SK SL SM SN SO SP SR SS ST SU SV SW SX SY
V	**Severn Valley**	Worcester	VA – VY

W	**West of England**	Exeter, Truro and Bristol	WA WB WC WD WE WF WG WH WJ WK WL WM WN WO WP WR WS WT WU WV WW WX WY
Y	**Yorkshire**	Leeds, Sheffield and Beverley	YA YB YC YD YE YF YG YH YJ YK YL YM YN YO YP YR YS YT YU YV YW YX YY

INTERNATIONAL DIALLING CODES

Country	Code	Country	Code
Afghanistan	93	Lithuania	370
Albania	355	Luxembourg	352
Algeria	213	Macau	853
American Samoa (1-)	684	Macedonia	389
Andorra	376	Madagascar	261
Angola	244	Malawi	265
Anguilla (1-)	264	Malaysia	60
Antarctica	672	Maldives	960
Antigua & Barbuda (1-)	268	Mali	223
Argentina	54	Malta	356
Armenia	374	Marshall Islands	692
Aruba	297	Mauritania	222
Australia	61	Mauritius	230
Austria	43	Mayotte	262
Azerbaijan	994	Mexico	52
Bahamas (1-)	242	Micronesia	691
Bahrain	973	Moldova	373
Bangladesh	880	Monaco	377
Barbados (1-)	246	Mongolia	976
Belarus	375	Montenegro	382
Belgium	32	Montserrat (1-)	664
Belize	501	Morocco	212

Benin	229	Mozambique	258
Bermuda (1-)	441	Myanmar	95
Bhutan	975	Namibia	264
Bolivia	591	Nauru	674
Bosnia & Herzegovina	387	Nepal	977
Botswana	267	Netherlands	31
Brazil	55	Neth Antilles	599
British Indian Ocean Terr.	246	New Caledonia	687
British Virgin Is (1-)	284	New Zealand	64
Brunei	673	Nicaragua	505
Bulgaria	359	Niger	227
Burkina Faso	226	Nigeria	234
Burundi	257	Niue	683
Cambodia	855	North Korea	850
Cameroon	237	N Mariana Is (1-)	670
Canada	1	Norway	47
Cape Verde	238	Oman	968
Cayman Is (1-)	345	Pakistan	92
Central African Rep	236	Palau	680
Chad	235	Palestine	970
Chile	56	Panama	507
China	86	Papua New Guinea	675
Christmas Island	61	Paraguay	595
Cocos Islands	61	Peru	51
Colombia	57	Philippines	63
Comoros	269	Pitcairn	64
Cook Islands	682	Poland	48
Costa Rica	506	Portugal	351

Croatia	385	Puerto Rico (1-)	939
Cuba	53	Qatar	974
Curacao	599	Rep of the Congo	242
Cyprus	357	Reunion	262
Czech Republic	420	Romania	40
Dem Rep of the Congo	243	Russia	7
Denmark	45	Rwanda	250
Djibouti	253	St Barthelemy	590
Dominica (1-)	767	St Helena	290
Dominican Rep (1-)	849	St Kitts & Nevis (1-)	869
East Timor	670	St Lucia (1-)	758
Ecuador	593	St Martin	590
Egypt	20	St P and Miquelon	508
El Salvador	503	St Vt + Gren's (1-)	784
Equatorial Guinea	240	Samoa	685
Eritrea	291	San Marino	378
Estonia	372	Sao T & Principe	239
Ethiopia	251	Saudi Arabia	966
Falkland Islands	500	Senegal	221
Faroe Islands	298	Serbia	381
Fiji	679	Seychelles	248
Finland	358	Sierra Leone	232
France	33	Singapore	65
French Polynesia	689	St Maarten (1-)	721
Gabon	241	Slovakia	421
Gambia	220	Slovenia	386
Georgia	995	Solomon Islands	677
Germany	49	Somalia	252
Ghana	233	South Africa	27

Gibraltar	350	South Korea	82
Greece	30	South Sudan	211
Greenland	299	Spain	34
Grenada (1-)	473	Sri Lanka	94
Guam (1-)	671	Sudan	249
Guatemala	502	Suriname	597
Guernsey (+44)	1481	S'bard & Jan Mayen	47
Guinea	224	Swaziland	268
Guinea-Bissau	245	Sweden	46
Guyana	592	Switzerland	41
Haiti	509	Syria	963
Honduras	504	Taiwan	886
Hong Kong	852	Tajikistan	992
Hungary	36	Tanzania	255
Iceland	354	Thailand	66
India	91	Togo	228
Indonesia	62	Tokelau	690
Iran	98	Tonga	676
Iraq	964	T'd & Tobago (1-)	868
Ireland	353	Tunisia	216
Isle of Man (+44)	1624	Turkey	90
Israel	972	Turkmenistan	993
Italy	39	Turks & C's Is (1-)	649
Ivory Coast	225	Tuvalu	688
Jamaica (1-)	876	U.S. Virgin Is (1-)	340
Japan	81	Uganda	256
Jersey (+44)	1534	Ukraine	380
Jordan	962	UAE	971
Kazakhstan	7	United Kingdom	44
Kenya	254	United States	1

Kiribati	686	Uruguay	598
Kosovo	383	Uzbekistan	998
Kuwait	965	Vanuatu	678
Kyrgyzstan	996	Vatican	379
Laos	856	Venezuela	58
Latvia	371	Vietnam	84
Lebanon	961	Wallis and Futuna	681
Lesotho	266	Western Sahara	212
Liberia	231	Yemen	967
Libya	218	Zambia	260
Liechtenstein	423	Zimbabwe	263

OTHER NUMERIC SYSTEMS

Each written language has its own set of depictions for numbers. These are from: https://creativecommons.org/ to which a donation has been made. There are many others.

Babylonian

1	𒁹	11	𒌋𒁹	21	𒎙𒁹	31	𒌍𒁹	41	𒅄𒁹	51	𒐐𒁹
2	𒈫	12	𒌋𒈫	22	𒎙𒈫	32	𒌍𒈫	42	𒅄𒈫	52	𒐐𒈫
3	𒐈	13	𒌋𒐈	23	𒎙𒐈	33	𒌍𒐈	43	𒅄𒐈	53	𒐐𒐈
4	𒐉	14	𒌋𒐉	24	𒎙𒐉	34	𒌍𒐉	44	𒅄𒐉	54	𒐐𒐉
5	𒐊	15	𒌋𒐊	25	𒎙𒐊	35	𒌍𒐊	45	𒅄𒐊	55	𒐐𒐊
6	𒐋	16	𒌋𒐋	26	𒎙𒐋	36	𒌍𒐋	46	𒅄𒐋	56	𒐐𒐋
7	𒐌	17	𒌋𒐌	27	𒎙𒐌	37	𒌍𒐌	47	𒅄𒐌	57	𒐐𒐌
8	𒐍	18	𒌋𒐍	28	𒎙𒐍	38	𒌍𒐍	48	𒅄𒐍	58	𒐐𒐍
9	𒐎	19	𒌋𒐎	29	𒎙𒐎	39	𒌍𒐎	49	𒅄𒐎	59	𒐐𒐎
10	𒌋	20	𒎙	30	𒌍	40	𒅄	50	𒐐		

Mayan

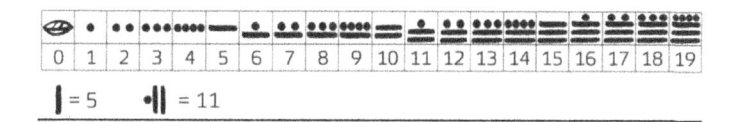

0	1	2	3	4	5	6	7	8	9	10	11	12	13	14	15	16	17	18	19

| = 5   •|| = 11

Indian

European	0	1	2	3	4	5	6	7	8	9
Arabic-Indic	.	١	٢	٣	٤	٥	٦	٧	٨	٩
Eastern Arabic-Indic (Persian and Urdu)	.	١	٢	٣	۴	۵	۶	٧	٨	٩
Devanagari (Hindi)	०	१	२	३	४	५	६	७	८	९
Tamil		க	உ	௩	௪	௫	௬	௭	௮	௯
Telugu	౦	౧	౨	౩	౪	౫	౬	౭	౮	౯

Egyptian

|	1 numeral	Stroke
∩	10 numeral	Fetter
૯	100 numeral	Coli
𓋹	1000 numeral	Lotus plant
𓂭	10,1000 numeral	finger
𓆏	100,000 numeral	tadpole
𓁨	1000,000 numeral	Heh god
⟵	One	Harpoon
⇓	Two	arrowhead
\\\\	Dual	two stokes
| | |	Plural	three strokes

APPENDIX A

SONGS WITH NUMBERS IN THE TITLE

Zero
Back to Zero .... Rolling Stones

One
Another One Bites the Dust .... Queen
(Call Me) Number One .... Tremeloes
Chant No 1 (I Don't Need this Pressure on) .... Spandau Ballet
Every 1's a Winner .... Hot Chocolate
Just One Look .... Hollies
Let Me be the One .... Shadows
Love in the First Degree .... Bananarama, S.A.W.
Love Plus One .... Haircut 100
Numero Uno .... Starlight
One U2
One .... Metallica
One Day at a Time .... Lena Martell
One Day I'll Fly Away .... Randy Crawford
One Day in Your Life .... Michael Jackson
One Fine Day .... Chiffons
One Goodbye in Ten .... Shara Nelson
One in Ten .... UB 40 / 808 State
One (is the Loneliest Number) .... Three Dog Night
One Love .... Bob Marley
One Love .... Prodigy
One Love .... Stone Roses
One Man Band .... Leo Sayer
....Baby One More Time .... Britney Spears
One Nation Under a Groove .... Funkadelic

One Night in Heaven	M People
One of These Days	Pink Floyd
One of These Nights	Eagles
One of Us	Abba
One Piece at a Time	Johnny Cash
One Two Three O'Leary	Des O'Connor
One Way or Another	Blondie
One Woman	Jade
Only One Woman	Marbles
Say I'm your No.1	Princess
You are the One	A-Ha

Two

2 Minutes to Midnight	Iron Maiden
Ain't 2 Proud 2 Beg	TLC
Goody Two Shoes	Adam Ant
It Takes Two	Tina Turner and Rod Stewart
Nothing Compares 2 You	Sinead O'Connor
Love Me Two Times	The Doors
One Two Three O'Leary	Des O'Connor
Song 2	Blur
The Two of Us	Mac and Katie Kissoon
Two Hearts	Phil Collins
Two Hearts	Bruce Springsteen
Two Little Boys	Rolf Harris
Two of Us	Beatles
Two Out of Three Ain't Bad	Meat Loaf
Two Pints of Lager and a Packet of Crisps Please	Splogenessa Bounds
Two Princes	Spin Doctors
Two Steps Behind	Def Leppard
Two Tribes	Frankie Goes to Hollywood
Two's Up	AC/DC
The Green Manalishi (with the Two Pronged Crown)	Fleetwood Mac

Three

3 x 3	Genesis
A Quarter to Three	Gary U. S. Bonds
Knock Three Times	Tony Orlando and Dawn
One Two Three O'Leary	Des O'Connor
The Three Bells	The Browns
Three Babies	Sinead O'Connor
Three Coins in the Fountain	Frank Sinatra
Three Little Birds	Bob Marley
Three Little Pigs	Green Jelly
Threepenny Opera Theme	Louis Armstrong/Billy Vaughn/Dick Hyman Trio
Three Steps to Heaven	Eddie Cochran
Three Times a Lady	Commodores

Four

409	Beach Boys
Four Horsemen	Metallica
Four Letter Word	Kim Wilde
Four Seasons in One Day	Crowded House
Four Sticks	Led Zeppelin
It's Four in the Morning	Faron Young
Luv 4 Luv	Robin S
Mary of the Fourth Form	Boomtown Rats
Positively 4th Street	Bob Dylan

Five

5.15	The Who
Five Live	George Michael & Queen
Five to One	The Doors
Five Years	David Bowie
Highway 5	Blessing
Just Who is the Five o'clock Hero	Jam
Rollin' in my 5.0	Vanilla Ice
Take Five	Dave Brubeck Quartet
Take 5	Northside

Six

Big Six	Judge Dread
I Got the Six	ZZ Top
Sixth Sense	Latino Rave
The Six Teens	Sweet

Seven

7	Prince
7 Ways to Love	Cola Boy
Big Seven	Judge Dread
Cloud Lucky Seven	Guy Mitchell
Return of the Los Palmas 7	Madness
Sailing on the Seven Seas	OMD
Seven	James
Seven Bridges Road	Eagles
Seven Days	Sting
Seven Drunken Nights	Dubliners
Seven Little Girls Sitting in the Back Seat	Avons/Bombalurina
Seven Nation Army	The White Stripes
Seven Rooms of Gloom	Four Tops
Seven Seas of Rhye	Queen
Seven Tears	Goombay Dance Band
Seven Wonders	Fleetwood Mac
Seventh Son of a Seventh Son	Iron Maiden

Eight

8 Days a Week	Beatles
Activ 8 (Come With Me)	Altern 8
Big Eight	Judge Dread
Driver 8	R.E.M.
Eight Day	Hazel O'Connor
Eight Miles High	The Byrds
Evapor 8	Altern 8
Hypnotis St 8	Altern 8
I'm Henry the Eighth	Herman's Hermits

Nine

#9 Dream	John Lennon and the Plastic Ono Band
9 to 5	Dolly Parton
9 to 5	Sheena Easton
Cloud Nine	Temptations
Cloud Nine	George Harrison
Love Potion Number Nine	Searchers
Nine Times out of Ten	Cliff Richard

Ten

11.59	Blondie
Big Ten	Judge Dread
Big Ten Inch Record	Aerosmith
I Close My Eyes and Count to Ten	Dusty Springfield
One in Ten	UB 40
Ten Years Gone	Led Zepellin

Teens

7teen	Regents
15 Years	Levellers
16 Bars	Stylistics
18 and Life	Skid Row
19	Paul Hardcastle
19th Nervous Breakdown	Rolling Stones
Edge of Seventeen	Stevie Nicks
Eighteen with a Ballet	Pete Wingfield
Hangar 18	Megedeth
Happy Birthday Sweet Sixteen	Neil Sedaka
Hey Nineteen	Steely Dan
My 16th Apology	Shakespear's Sister
Only Sixteen	Craig Douglas
Seventeen	Sex Pistols
Sixteen Candles	The Crests
Sixteen Tons	Tennessee Ernie Ford
Sweet Little Sixteen	Chuck Berry
TVC 15	David Bowie

When You were Sweet Sixteen .... Fureys and Davey Arthur
You're Sixteen................................. Johnny Burnette / Ringo Starr

Twenties
20 Seconds to Comply ....Silver Bullet
21 Guns .... Green Day
24 Hours... Betty Boo
25 Minutes to Go .... Johnny Cash
25 or 6 to 4 ....Chicago
29 Palms ....Robert Plant
Twentieth Century Boy ....T. Rex
Twenty Four Hours from Tulsa.................................... Gene Pitney
Twenty Tiny Fingers... Stargazers

Thirties
32°F Boilerhouse... Pop Will Eat Itself
'39 .... Queen
36D .... Beautiful South

Forties
40 Miles...Congress
48 Crash .... Susie Quatro
48 Hours..The Clash
Forty Miles of Bad Road...Duane Eddy
Stars on 45 .... Starsound

Fifties
57 Channels...Bruce Springsteen
Fifty Ways to Leave Your Lover .... Paul Simon
The 59th Street Bridge Song ....Simon and Garfunkel

Sixties
Car 67.. Driver 67
December '63 (Oh What a Night) .... Four Seasons
Highway 61 Revisited ....Bob Dylan
Route 66 ....Nat King Cole

Summer '68 ....Pink Floyd
Summer of '69 ....Bryan Adams
When I'm 64..Beatles

Nineties
96 Tears .... Stranglers
99 Red Balloons.. Nena
99 Problems ....Jay-Z
Ninety Nine Ways .... Tab Hunter

Hundreds etc.
100 miles and Runnin' ....NWA
100% .... Sonic Youth
137 Disco Heaven...Amii Stewart
747 (Strangers in the Night)... Saxon
911 is a joke... Public Enemy
™231 .... Anticapella
A Hundred Pounds of Clay......................................Craig Douglas
Fields of Fire (400 miles)... Big Country
I'm Gonna Be (500 miles)...Proclaimers
409 ....Beach Boys

Thousands
1921 ....The Who
1984...David Bowie
1999...Prince
2000 Man.. Rolling Stones
2000 Miles..Pretenders
2112 ....Rush
100,000 Years .... Kiss
Land of 1000 Dances..Wilson Pickett
Swords of a Thousand Men ....Ten Pole Tudor
The Night has a Thousand Eyes .... Bobby Vee

Millions etc.
A Million Love Songs ....Take That
Billion Dollar Babies..Alice Cooper
Infinity ....Guru Josh

Multiples
1-2-3 ....Len Barry, Gloria Estefan & the Miami Sound Machine
2-4-6-8 Motorway.. Tom Robinson Band
4-3-2-1 .... Manfred Mann
5-7-0-5 ....City Boy
9 times out of 10 ....Cliff Richard
9 to 5 .... Sheena Easton
25 or 6 to 4 ....Chicago
Rainy Day Women Nos. 12 & 35 ....Bob Dylan

APPENDIX B
FILMS WITH NUMBERS IN THE TITLE

Zero

Zero Effect..............(1998) Bill Pullman, Ben Stiller, Ryan O'Neal
Patient Zero ....(2018) Matt Smith, Natalie Dormer, Clive Standen
Less Than Zero..................(1987) Andrew McCarthy, Jami Gertz, Robert Downey Jr.
Zero Dark Thirty ....(2012) Jessica Chastain, Joel Edgerton, Chris Pratt
The Zero Theorem ....(2013) Christoph Waltz, Lucas Hedges, Mélanie Thierry
Horizon: Zero Dawn ....(2017) Ashly Burch, Laura van Tol, Ava Potter

One

First ....(2012) John Orozco, Heena Sidhu, David Rudisha
The One ....(2001) Jet Li, Carla Gugino, Jason Statham
One Day ...(2011) Anne Hathaway, Jim Sturgess, Patricia Clarkson
First Kill ......(2017) Bruce Willis, Hayden Christensen, Ty Shelton
First Man ....(2018) Claire Foy, Jon Bernthal, Ryan Gosling
Year One................(2009) Jack Black, Michael Cera, Olivia Wilde
One Week...............(1920) Buster Keaton, Sybil Seely, Joe Roberts
One Week.......(2008) Joshua Jackson, Peter Spence, Marc Strange
First Knight ....(1995) Sean Connery, Richard Gere, Julia Ormond
One Fine Day ....(1996) Michelle Pfeiffer, George Clooney, Mae Whitman
Air Force One ..(1997) Harrison Ford, Gary Oldman, Glenn Close
The First Time..................(2012) Dylan O'Brien, Britt Robertson, Victoria Justice

Capricorn One.........................(1977) Elliott Gould, James Brolin, Brenda Vaccaro
One Among Us.......(2005) James Russo, Christean Nahas, Brandt Wille
One Way Ticket.. [numerous]
One More Time................(1970) Sammy Davis Jr., Peter Lawford, John Wood
One More Time.......... (2015) Christopher Walken, Amber Heard, Kelli Garner
One False Move................(1992) Bill Paxton, Billy Bob Thornton, Cynda Williams
One Night Stand ....(1997) Wesley Snipes, Nastassja Kinski, Kyle MacLachlan
One, Two, Three ....(1961) James Cagney, Horst Buchholz, Pamela Tiffin
One Eight Seven ....(1997) Samuel L Jackson, John Heard, Kelly Rowan
The Big Red One...................... (1980) Lee Marvin, Mark Hamill, Robert Carradine
One Hour Photo............. (2002) Robin Williams, Connie Nielsen, Michael Vartan
X-Men First Class .... (2011) James McAvoy, Michael Fassbender
One Way Passage .... (1932) William Powell, Kay Francis, Frank McHugh
Ready Player One.................... (2018) Tye Sheridan, Olivia Cooke, Hannah John-Kamen
The 1 Second Film ....(2017)
One Night of Love............ (1934) Grace Moore, Tullio Carminati, Lyle Talbot
One for the Money .... (2012) Kathrine Heigl, Jason O'Mara, Daniel Sunjata
Take One False Step .... (1949) William Powell, Shelley Winters, Marsha Hunt
It Happened One Night..... (1934) Clark Gable, Claudette Colbert
Earth: One Amazing Day...... (2017) Robert Redford, Jackie Chan

One Night With the King...... (2006) Tiffany Dupont, Luke Goss, John Noble
Rogue One: A Star Wars Story... (2016) Felicity Jones, Diego Luna
One of Our Dinosaurs is Missing .... (1975) Peter Ustinov, Helen Hayes
The One That Got Away ...... (1957) Hardy Krüger, Colin Gordon, Michael Goodliffe
One Flew Over the Cuckoo's Nest .... (1974) Jack Nicholson, Louise Fletcher
Captain America: The First Avenger .... (2011) Chris Evans, Hugo Weaving
The Taking of Pelham One Two Three ..... (1974) Walter Matthau, Robert Shaw

Two

Two Lovers .... (2008) Joaquin Phoenix, Gwyneth Paltrow, Vinessa Shaw
Two Women .... (1960) Sophia Loren, Jean-Paul Belmondo, Raf Vallone
Two Women .... (2014) Ralph Fiennes, Sylvie Testud, Aleksandr Baluev
Second Sight .... (1989) John Larroquette, Bronson Pinchot, Bess Armstrong
Second Nature .... (2016) Collette Wolfe, Sam Huntington, Carollani Sandberg
Two Solitudes .... (1978) Jean-Pierre Aumont, Stacy Keach, Gloria Carlin
The Two Jakes .... (1990) Jack Nicholson, Harvey Keitel, Meg Tilly
2 Fast 2 Furious..... (2003) Paul Walker, Tyrese Gibson, Cole Hauser
Two for the Road .... (1967) Audrey Hepburn, Albert Finney, Eleanor Bron
Two Weeks Notice .... (2002) Sandra Bullock. Hugh Grant, Alicia Witt
Two for the Money ..... (2005) Matthew McConaughey, Al Pacino, Rene Russo

Two Mules for Sister Sara .... (1970) Clint Eastwood, Shirley MacLaine
Terminator2: Judgement Day .... (1991)Arnold Schwarzeneggar, Linda Hamilton
Lord of the Rings: The Two Towers................(2002) Elijah Wood, Ian McKellen
Lock, Stock and Two Smoking Barrels .... (1998) Jason Statham, Vinnie Jones
The Second Best Exotic Marigold Hotel........... (2015) Judi Dench, Maggie Smith

Three
3 Ninjas .... (1992) Victor Wong, Michael Treanor, Max Elliott Slade
Life of Pi ...... (2012) Suraj Sharma, Irrfan Khan, Adil Hussain, Tiger
Three Kings ..... (1999) George Clooney, Mark Wahlberg, Ice Cube
Third Person .... (2013) Liam Neeson, Mila Kunis, Adrien Brody
3 Ring Circus.......... (1954) Dean Martin, Jerry Lewis, Joanne Dru
Three Sisters ...... (1970) Jeanne Watts, Joan Plowright, Louise Purnell
Three Sisters ..... (1988) Fanny Ardant, Greta Scacchi, Valeria Golino
Three Sisters .... (2001) Katherine LaNasa, Dyan Cannon, David Alan Basche
Three Burials .... (2005) Tommy Lee Jones, Barry Pepper, Dwight Yoakam
Three Amigos .... (1986) Steve Martin, Chevy Chase, Martin Short
The Third Man ... (1949) Orson Welles, Joseph Cotten, Alida Valli
Three to Tango .... (1999) Neve Campbell, Matthew Perry, Dylan McDermott
Shrek the Third .... (2007) Mike Myers, Cameron Diaz, Eddie Murphy
3 Way Junction .... (2017) Tom Sturridge, Tommy Flanagan, Stacy Martin
Table for Three .... (2009) Brandon Routh, Jesse Bradford, Ed Ackerman
Three Fugitives .... (1989) Nick Nolte, Martin Short, Sarah Rowland Doroff

Three Comrades ....(1938) Robert Taylor, Margaret Sullavan,
Franchot Tone
Three on a Couch .... (1966) Jerry Lewis, Janet Leigh,
Mary Ann Mobley
Three on a Match .... (1932) Joan Blondell, Warren William,
Ann Dvorak
3 Ninjas Kick Back .... (1994) Victor Wong, Max Elliott Slade,
Sean Fox
The Three Stooges .... (2000) Paul Ben-Victor, Evan Handler,
John Kassir
The Three Stooges ......(2012) Sean Hayes, Chris Diamantopoulos,
Will Sasso
Three Colors: Red........ (1994) Irène Jacob, Jean-Louis Trintignant
Three Colors: Blue.................................... (1993) Juliette Binoche,
Zbigniew Zamachowski
Three Colors: White ...... (1994) Zbigniew Zamachowski, Julie Delpy
3 Men and a Baby.............. (1987) Tom Selleck, Steve Guttenberg,
Ted Danson
Three for the Road (1987) Charlie Sheen, Kerri Green, Alan Ruck
Three Young Texans .... (1954) Mitzi Gaynor, Jeffrey Hunter,
Keefe Brasselle
The Next Three Days .... (2010) Russell Crowe, Elizabeth Banks,
Liam Neeson
The Three Caballeros ....(1944) Aurora Miranda, Carmen Molina,
Dora Luz
Three the Hard Way.............. (1974) Jim Brown, Fred Williamson,
Jim Kelly
The Three Musketeers .... (1948) Lana Turner, Gene Kelly,
June Allyson
The Three Musketeers .... (1973) Oliver Reed, Raquel Welch
The Three Musketeers ..... (1993) Charlie Sheen, Kiefer Sutherland
The Three Musketeers.....(2011) Logan Lerman, Matthew Macfadyen
3,2,1…Frankie Go Boom.....(2012) Charlie Hunnam, Chris O'Dowd
3 Men and a Little Lady.......(1990) Tom Selleck, Steve Guttenberg
The Three Faces of Eve...... (1957) Joanne Woodward, Lee J. Cobb
Three Days of the Condor ..... (1975) Robert Redford, Faye Dunaway

Three Coins in the Fountain .... (1954) Clifton Webb, Dorothy McGuire
Close Encounters of the Third Kind .... (1977) Richard Dreyfuss
Terminator 3: Rise of the Machines ..... (2003) Arnold Schwarzenegger
Snow White and the Three Stooges .... (1961) Moe Howard, Larry Fine
Three Billboards Outside Ebbing, Missouri .... (2017) Frances McDormand

Four

4D Man ...... (1959) Robert Lansing, Lee Meriwether, James Congdon
4 For Texas .... (1963) Frank Sinatra, Dean Martin, Anita Ekberg
Four Rooms .... (1995) Tim Roth, Antonio Banderas, Sammi Davis
Fourth Floor .... (2003) Linden Banks, Guy Fletcher, Jd Mendonca
Four Brothers .... (2005) Mark Wahlberg, Tyrese Gibson, André Benjamin
Fantastic Four .... (2005) Ioan Gruffudd, Michael Chiklis, Chris Evans
Fantastic Four ...... (2015) Miles Teller, Kate Mara, Michael B. Jordan
4 Minute Mile ...... (2014) Kelly Blatz, Richard Jenkins, Kim Basinger
The Fourth Kind ...... (2009) Milla Jovovich, Elias Koteas, Will Patton
Four Daughters .... (1938) Claude Rains, John Garfield, Jeffrey Lynn
The Four Seasons ..... (1981) Alan Alda, Carol Burnett, Len Cariou
The Four Feathers .... (1929) Richard Arlen, Fay Wray, Clive Brook
The Four Feathers .... (1939) John Clements, Ralph Richardson
The Four Feathers .... (1978) Beau Bridges, Robert Powell, Simon Ward
The Four Feathers .... (2002) Heath Ledger, Wes Bentley, Kate Hudson
The Fantastic Four .... (1994) Alex Hyde-White, Jay Underwood
I am Number Four .... (2011) Alex Pettyfer, Timothy Olyphant, Dianna Agron
The Fourth Protocol .... (1987) Michael Caine, Pierce Brosnan, Ned Beatty
The Fourth Wise Man .... (1985) Martin Sheen, Alan Arkin, Eileen Brennan

The Fourth Dimension .... (2012) Val Kilmer, Rachel Korine,
Igor Sergeev
Born on the Fourth of July ....(1989) Tom Cruise, Raymond J. Barry
Four Weddings and a Funeral......................... (1994) Hugh Grant,
Andie MacDowell
Four Months, 3 Weeks, 2 Days............ (2007) Anamaria Marinca,
Laura Vasiliu
Fantastic Four: Rise of the Silver Surfer...... (2007) Ioan Gruffudd,
Jessica Alba
The Four Musketeers: Milady's Revenge........ (1974) Michael York,
Raquel Welch
The Four Horsemen of the Apocalypse...... (1921) Rudolph Valentino
The Four Horsemen of the Apocalypse.............(1962) Glenn Ford,
Ingrid Thulin

Five

5 Fingers......(1952) James Mason, Danielle Darrieux, Michael Rennie
5 Card Stud...... (1968) Dean Martin, Robert Mitchum, Inger Stevens
Fifth Ave Girl....................(1939) Ginger Rogers, Walter Connolly,
Verree Teasdale
Five Easy Pieces .... (1970) Jack Nicholson, Karen Black,
Billy Green Bush
The Fifth Estate........(2013) Benedict Cumberbatch, Daniel Brühl
The Fifth Element .... (1997) Bruce Willis, Milla Jovovich,
Gary Oldman
Five Graves to Cairo .... (1943) Franchot Tone, Anne Baxter,
Akim Tamiroff
Five Weeks in a Balloon.....(1962) Red Buttons, Fabian, Barbara Eden
Slaughterhouse – Five .... (1972) Michael Sacks, Ron Leibman,
Eugene Roche
Five Minutes of Heaven ....(2009) Liam Neeson, James Nesbitt
The Five-Year Engagement .... (2012) Jason Segel, Emily Blunt,
Chris Pratt

Six

6 Days...................(2017) Jamie Bell, Mark Strong, Abbie Cornish

The 6th Day...... (2000) Arnold Schwarzenegger, Michael Rapaport
The Sixth Man .... (1997) Marlon Wayans, Kadeem Hardison,
David Paymer
Sixth and Main..................(1977) Leslie Nielsen, Beverly Garland,
Roddy McDowall
The Sixth Sense ....(1999) Bruce Willis, Haley Joel Osment,
Toni Collette
Six Days Seven Nights .... (1998) Harrison Ford, Anne Heche
Six Degrees of Separation .... (1993) Will Smith, Stockard Channing
The Inn of the Sixth Happiness.................. (1958) Ingrid Bergman,
Robert Donat

Seven
Se7en................. (1995) Morgan Freeman, Brad Pitt, Kevin Spacey
7 Women ..... (1966) Anne Bancroft, Sue Lyon, Margaret Leighton
7th Heaven............. (1927) Janet Gaynor, Charles Farrell, Ben Bard
Seventh Heaven .... (1937) Simone Simon, James Stewart,
Jean Hersholt
7 Plus Seven .... (1970) Bruce Balden, Jacqueline Bassett,
Symon Basterfield
Seventh Son .... (2014) Ben Barnes, Julianne Moore, Jeff Bridges
Seven Pounds........................ (2008) Will Smith, Rosario Dawson,
Woody Harrelson
Seven Samurai .... (1954) Toshirô Mifune, Takashi Shimura,
Keiko Tsushima
Seven Beauties .... (1976) Giancarlo Giannini, Fernando Rey,
Shirley Stoler
The Seventh Sign....................(1988) Demi Moore, Michael Biehn,
Jürgen Prochnow
The Seventh Seal ......(1957) Max von Sydow, Gunnar Björnstrand
Seven Psychopaths ....(2012) Colin Farrell, Woody Harrelson,
Sam Rockwell
The Seven Year Itch .... (1955) Marilyn Monroe, Tom Ewell,
Evelyn Keyes
Seven Years in Tibet ......(1997) Brad Pitt, David Thewlis, BD Wong

The Magnificent Seven......... (1960 Yul Brynner, Charles Bronson, Steve McQueen
The Magnificent Seven..... (2016) Denzel Washington, Chris Pratt
Seven Brides for Seven Brothers .... (1954) Jane Powell, Howard Keel
Snow White and the Seven Dwarfs.......... (1937) Adriana Caselotti

Eight
8 ½ ....(1963) Marcello Mastroianni, Anouk Aimée, Claudia Cardinale
8mm ......(1999) Nicolas Cage, Joaquin Phoenix, James Gandolfini
8 Mile .... (2002) Eminem, Brittany Murphy, Kim Basinger
Super 8.............. (2011) Elle Fanning, AJ Michalka, Kyle Chandler
8 Seconds......(1994) Luke Perry, Stephen Baldwin, James Rebhorn
8 Women .... (2002) Fanny Ardant, Emmanuelle Béart, Danielle Darrieux
Jennifer 8...... (1992) Andy Garcia, Uma Thurman, Lance Henriksen
Eight Below....... (2006) Paul Walker, Jason Biggs, Bruce Greenwood
BUtterfield 8.................(1960) Elizabeth Taylor, Laurence Harvey, Eddie Fisher
Ocean's Eight.................... (2018) Sandra Bullock, Cate Blanchett, Anne Hathaway
Eight Men Out .... (1988) John Cusack, Clifton James, Michael Lerner
Eight on the Lam......................... (1967) Bob Hope, Phyllis Diller, Jonathan Winters
The Hateful Eight..............(2015) Samuel L. Jackson, Kurt Russell
Eight Crazy Nights .... (2002) Adam Sandler, Rob Schneider, Jackie Sandler
Eight Legged Freaks............. (2002) David Arquette, Kari Wuhrer, Scott Terra
When Eight Bells Toll...... (1971) Anthony Hopkins, Jack Hawkins
8 Heads in a Duffel Bag............... (1997) Joe Pesci, Andy Comeau, Kristy Swanson
The Adventures of Buckaroo Banzai Across the 8th Dimension....... ....Peter Weller

Nine
9/11 .... (2017) Charlie Sheen, Gina Gershon, Whoopi Goldberg
9 Songs .... (2004) Kieran O'Brien, Margo Stilley,
District 9 .... (2009) Sharlto Copley, David James, Jason Cope
Session 9 ...... (2001) David Caruso, Stephen Gevedon, Paul Guilfoyle
9½ weeks .... (1986) Mickey Rourke, Kim Basinger,
Margaret Whitton
9 to 5 .... (1980) Jane Fonda, Lily Tomlin, Dolly Parton
Nine Months .... (1995) Hugh Grant, Julianne Moore, Tom Arnold
The Ninth Gate ...... (1999) Johnny Depp, Frank Langella, Lena Olin
The Whole Nine Yards .... (2000) Bruce Willis, Matthew Perry
The Ninth Configuration .... (1980) Stacy Keach, Scott Wilson,
Jason Miller

Ten
10 .... (1979) Dudley Moore, Bo Derek, Julie Andrews
The Ten .... (2007) Paul Rudd, Jessica Alba, Winona Ryder
10 Years ..... (2011) Channing Tatum, Rosario Dawson, Chris Pratt
Starter for 10 .... (2006) James McAvoy, Alice Eve, Rebecca Hall
The 10th Victim .... (1965) Marcello Mastroianni, Ursula Andress
10 Items or Less .... (2006) Morgan Freeman, Paz Vega, Jonah Hill
Ten Little Indians ..... (1965) Hugh O'Brian, Shirley Eaton, Fabian
Ten Little Indians .... (1974 Charles Aznavour, Maria Rohm,
Adolfo Celi
Ten Little Indians .... (1989) Donald Pleasence, Brenda Vaccaro,
Frank Stallone
10, Rillington Place ... (1971) Richard Attenborough, Judy Geeson,
John Hurt
10, Cloverfield Lane .... (2016) John Goodman,
Mary Elizabeth Winstead
Tenth Avenue Angel .... (1948) Margaret O'Brien, Angela Lansbury
The Whole Ten Yards .... (2004) Bruce Willis, Matthew Perry
The Ten Commandments .... (1923) Theodore Roberts,
Charles de Rochefort
The Ten Commandments .... (1956) Charlton Heston, Yul Brynner
The Ten Commandments .... (2007) Ben Kingsley, Christian Slater

Slaughter on 10th Avenue........... (1957) Richard Egan, Jan Sterling, Dan Duryea
10 Things I Hate About You........ (1999) Heath Ledger, Julia Stiles
10 Rules for Sleeping Around .... (2013) Tammin Sursok, Virginia Williams
How to Lose a Guy in 10 Days ....(2003) Kate Hudson, Matthew McConaughey
10 Questions for the Dalai Lama............. (2006) The Dalai Lama, Tenzin Bagdro
The Ten Commandments: The Movie ...(2016) Guilherme Winter

Eleven
Ocean's Eleven.......................(1960) Frank Sinatra, Dean Martin, Sammy Davis Jr.
Ocean's Eleven.... (2001) Brad Pitt, George Clooney, Matt Damon
The Eleventh Hour........... (2008) Matthew Reese, Jennifer Klekas
The Eleventh Hour Guest............ (1945) Jean Tissier, Roger Pigaut

Twelve
12 Strong .... (2018) Chris Hemsworth, Michael Shannon, Michael Peña
12 Rounds ....(2009) John Cena, Ashley Scott, Aidan Gillen
Twelfth Night .... (1980) Alec McCowen, Robert Hardy, Felicity Kendal
Twelfth Night .... (2013) Samuel Barnett, Liam Brennan, Paul Chahidi
Ocean's Twelve.... (2004) George Clooney, Brad Pitt, Julia Roberts
12 Years a Slave .... (2013) Chiwetel Ejiofor, Michael Kenneth Williams
The Dirty Dozen .... (1967) Lee Marvin, Ernest Borgnine, Telly Savalas
Twelve Monkeys .... (1995) Bruce Willis, Madeleine Stowe, Brad Pitt
The Twelve Chairs .... (1962) Mel Brooks, Ron Moody, Frank Langella
Twelve Angry Men.................... (1957) Henry Fonda, Lee J. Cobb, Martin Balsam

The Twelve Tasks of Asterix...... (1976) Roger Carel, Jacques Morel
Twelfth Night or What You Will.... (1996) Helena Bonham Carter

Thirteen
13 Hours........................... (2016) John Krasinski, Pablo Schreiber,
                                                        James Badge Dale
Thirteen..... (2003) Evan Rachel Wood, Holly Hunter, Nikki Reed
Apollo 13 .... (1995) Tom Hanks, Bill Paxton, Kevin Bacon
Thirteen Days............... (2000) Kevin Costner, Bruce Greenwood,
                                                         Shawn Driscoll
Friday the 13th .... (1980) Betsy Palmer, Adrienne King,
                                                          Jeannine Taylor
Friday the 13th .... (2009) Jared Padalecki, Amanda Righetti,
                                                            Derek Mears
13 Going on 30......(2004) Jennifer Garner, Mark Ruffalo, Judy Greer
Thirteen Ghosts......................(1960) Charles Herbert, Jo Morrow,
                                                           Martin Milner
The 13th Warrior............. (1999) Antonio Banderas, Diane Venora,
                                                          Dennis Storhøi
Ocean's Thirteen (2007) George Clooney, Brad Pitt, Matt Damon
The Thirteen Floor.................. (1999) Craig Bierko, Gretchen Mol
Assault of Precinct 13............(1976) Austin Stoker, Darwin Joston,
                                                          Laurie Zimmer
Assault on Precinct 13 .... (2005) Ethan Hawke, Laurence Fishburne
Friday the 13th:Part 2 .... (1981) Betsy Palmer, Amy Steel, John Furey
Friday the 13th:Part III .... (1982) Dana Kimmell, Tracie Savage
Friday the 13th:The Final Chapter.............. (1984) Erich Anderson,
                                                          Judie Aronson
Thirteen Conversations About One Thing....... (2001) Alan Arkin,
                                                          John Turturro

Fourteen
Fourteen Hours................. (1951) Paul Douglas, Richard Basehart,
                                                      Barbara Bel Geddes
14 Days to Life.................... (1997) Kai Wiesinger, Michael Mendl,
                                                      Katharina Meinecke

Fifteen
15 Minutes .... (2001) Robert De Niro, Edward Burns, Kelsey Grammer

Sixteen
16 Blocks .... (2006) Bruce Willis, Yasiin Bey, David Morse
Sixteen Candles .... (1984) Molly Ringwald, Anthony Michael Hall, Justin Henry

Seventeen
17 Again .... (2009) Zac Efron, Matthew Perry, Leslie Mann
Stalag 17 ...... (1953) William Holden, Don Taylor, Otto Preminger

Eighteen
18 Again! .... (1998) Charlie Schlatter, George Burns, Tony Roberts
Apollo 18 ...... (2011) Warren Christie, Lloyd Owen, Ryan Robbins
Eighteen .... (2005) Paul Anthony, Brendan Fletcher, Clarence Sponagle

Nineteen
Vehicle 19 ..... (2013) Paul Walker, Naima McLean, Gys de Villiers
Nineteen Nineteen .... (1985) Paul Scofield, Maria Schell, Frank Finlay
K-19: The Widowmaker (2002) Harrison Ford, Sam Spruell, Peter Stebbings

Twenties
20 Centimetres .... (2005) Mónica Cervera, Pablo Puyol, Miguel O'Dogherty
Halloween H20: 20 Years Later .... (1998) Jamie Lee Curtis, Josh Hartnett
21 .... (2008) Jim Sturgess, Kate Bosworth, Kevin Spacey
21 Grams .... (2003) Sean Penn, Benicio Del Toro, Naomi Watts
21 and Over .... (2013) Miles Teller, Justin Chon, Jonathan Keltz
21 Jump Street .... (1987) Johnny Depp, Dustin Nguyen, Peter DeLuise

21 Jump Street ....(2012) Jonah Hill, Channing Tatum, Ice Cube
Mile 22....... (2018) Mark Wahlberg, Lauren Cohan, John Malkovich
Catch 22........ (1970) Alan Arkin, Martin Balsam, Richard Benjamin
22 Bullets .... (2010) Jean Reno, Kad Merad, Jean-Pierre Darroussin
22 Jump Street.......(2014) Channing Tatum, Jonah Hill, Ice Cube
The Number 23 .... (2007) Jim Carrey, Virginia Madsen,
Logan Lerman
23 Paces to Baker Street............... (1956) Van Johnson, Vera Miles,
Cecil Parker
24 Hour Party People ... (2002) Steve Coogan, Lennie James, John
Thomson
The 25th Hour...................... (2002) Edward Norton, Barry Pepper
27 Dresses....................... (2008) Katherine Heigl, James Marsden,
Malin Akerman
28 Days. (2000) Sandra Bullock, Viggo Mortensen, Dominic West
28 Days Later................... (2002) Cillian Murphy, Naomie Harris,
Christopher Eccleston
28 Weeks Later ..... (2007) Jeremy Renner, Rose Byrne, Robert Carlyle
29 Palms .... (2002) Jeremy Davies, Chris O'Donnell,
Rachael Leigh Cook
29th Street ...... (1991) Anthony LaPaglia, Danny Aiello, Lainie Kazan
Track 29 .... (1988) Theresa Russell, Gary Oldman, Christopher Lloyd

Thirties

30 Days of Night ....(2007) Josh Hartnett, Melissa George,
Danny Huston
30 Minutes or Less........... (2011) Jesse Eisenberg, Danny McBride,
Nick Swardson
31......(2016) Malcolm McDowell, Richard Brake, Jeff Daniel Phillips
Naked Gun 33 1/3: The Final Insult .... (1994) Leslie Nielsen,
Priscilla Presley
Miracle on 34th Street....... (1947) Edmund Gwenn, Maureen O'Hara
Miracle on 34th Street .... (1973) Sebastian Cabot, Jane Alexander
Miracle on 34th Street .... (1994) Richard Attenborough,
Elizabeth Perkins
36 Hours.......... (1964) James Garner, Eva Marie Saint, Rod Taylor

36th Precinct................. (2004) Daniel Auteuil, Gérard Depardieu, André Dussollier
To Gillian on Her 37th Birthday.................(1996) Peter Gallagher, Michelle Pfeiffer
Thirty Nine ....(2016) Marshall Bell, Josh Evans, Natasha Gregson Wagner
Glorious 39 ..... (2009) Romola Garai, Eddie Redmayne, Juno Temple
The 39 Steps...................(1935) Robert Donat, Madeleine Carroll, Lucie Mannheim
The 39 Steps....... (1959) Kenneth More, Taina Elg, Brenda de Banzie
The 39 Steps....... (1978) Robert Powell, David Warner, Eric Porter
The 39 Steps .... (2008) Rupert Penry-Jones, Lydia Leonard, David Haig

Forties
This is 40.............(2012) Paul Rudd, Leslie Mann, Maude Apatow
The 40 Year Old Virgin ..... (2005) Steve Carell, Catherine Keener, Paul Rudd
40 Days and 40 Nights....... (2002) Josh Hartnett, Shannyn Sossamon
42 .... (2013) Chadwick Boseman, T.R. Knight, Harrison Ford
42nd Street .... (1933) Warner Baxter, Bebe Daniels, George Brent
Movie 43 .... (2013) Emma Stone, Stephen Merchant, Richard Gere
44 Minutes: The North Hollywood Shoot-Out .... (2003) Michael Madsen
.45 .... (2006) Milla Jovovich, Angus Macfadyen, Stephen Dorff
Love and a .45 .... (1994) Gil Bellows, Renée Zellweger, Rory Cochrane
47 Ronin.... (2013) Keanu Reeves, Hiroyuki Sanada, Ko Shibasaki
48 Hrs........... (1982) Nick Nolte, Eddie Murphy, Annette O'Toole
Ladder 49 .... (2004) Joaquin Phoenix, John Travolta, Jacinda Barrett
49th Parallel .... (1941) Leslie Howard, Laurence Olivier, Richard George

Fifties
52 Pick-up .... (1986) Roy Scheider, Ann-Margret, Vanity
54 .... (1998) Ryan Phillippe, Salma Hayek, Neve Campbell

50/50 .... (2011) Joseph Gordon-Levitt, Seth Rogen, Anna Kendrick
Planet 51 .... (2009) Dwayne Johnson, Seann William Scott, Jessica Biel
Formula 51.................... (2001) Samuel L. Jackson, Robert Carlyle, Emily Mortimer
Passenger 57 .... (1992) Wesley Snipes, Bruce Payne, Tom Sizemore
50 First Dates.................. (2004) Adam Sandler, Drew Barrymore, Rob Schneider
55 Days at Peking ....(1963) Charlton Heston, Ava Gardner, David Niven
Fifty Shades of Grey .... (2015) Dakota Johnson, Jamie Dornan, Jennifer Ehle
Fifty Shades Darker .... (2017) Dakota Johnson, Jamie Dornan, Eric Johnson
Fifty Shades Freed............. (2018) Dakota Johnson, Jamie Dornan, Arielle Kebbel

Sixties
Interstate 60: Episodes on the Road........... (2002) James Marsden, Gary Oldman
Gone in Sixty Seconds ....(2000) Nicolas Cage, Angelina Jolie, Giovanni Ribisi
61* ...... (2001) Barry Pepper, Thomas Jane, Anthony Michael Hall
Buffalo '66..... (1998) Vincent Gallo, Christina Ricci, Ben Gazzara

Seventies
'71.......................... (2014) Jack O'Connell, Sam Reid, Sean Harris
Winchester '73................... (1950) James Stewart, Shelley Winters, Dan Duryea

Eighties
Around the World in 80 Days....... (1956) David Niven, Cantinflas
Around the World in 80 Days (2004) Jackie Chan, Steve Coogan, Jim Broadbent
84, Charing Cross Road ......(1987) Anne Bancroft, Anthony Hopkins
88 Minutes...............(2007) Al Pacino, Alicia Witt, Ben McKenzie

Nineties
United 93 .... (2006) David Alan Basche, Olivia Thirlby, Liza Colón-Zayas
99 Homes ..... (2014) Andrew Garfield, Michael Shannon, Laura Dern

One hundreds
100 Girls .... (2000) Jonathan Tucker, Emmanuelle Chriqui, James DeBello
100 Rifles .... (1969) Jim Brown, Raquel Welch, Burt Reynolds
127 Hours .... (2010) James Franco, Amber Tamblyn, Kate Mara
One Eight Seven .... (1997) James Franco, Amber Tamblyn, Kate Mara
The Taking of Pelham 123 .... (2009) Denzil Washington, John Travolta
One Hundred Men and a Girl .... (1937) Deanna Durbin, Adolphe Menjou
One Hundred and One Dalmatians .... (1961) Rod Taylor, Betty Lou Gerson
One Hundred and One Dalmatians .... (1996) Glenn Close, Jeff Daniels
One Hundred and Two Dalmatians .... (2000) Glenn Close, Gérard Depardieu

Two hundreds
200 Cigarettes ...... (1999) Ben Affleck, Casey Affleck, Dave Chappelle
RKO 281 .... (1999) Liev Schreiber, James Cromwell, Melanie Griffith
#211 .... (2018) Nicolas Cage, Sophie Skelton, Michael Rainey Jr.

Three hundreds
300 .... (2006) Gerard Butler, Lena Headey, David Wenham
The 300 Spartans .... (1962) Richard Egan, Ralph Richardson, Diane Baker
300 Rise of an Empire .... (2013) Sullivan Stapleton, Eva Green, Lena Headey
360 .... (2011) Rachel Weisz, Jude Law, Anthony Hopkins

Four Hundreds
The 400 Blows...................(1959) Jean-Pierre Léaud, Albert Rémy, Claire Maurier
Dimension 404..... (2017) Mark Hamill, Catherine Farrington Garcia
Fahrenheit 451 ....(1966) Oskar Werner, Julie Christie, Cyril Cusack
Fahrenheit 451 .... (2018) Sofia Boutella, Michael Shannon, Michael B. Jordan

Five hundreds
(500) Days of Summer............................ (2009) Zooey Deschanel, Joseph Gordon-Levitt
U-571 ..... (2000) Matthew McConaughey, Bill Paxton, Harvey Keitel

Six hundreds
633 Squadron.................. (1964) Cliff Robertson, George Chakiris, Maria Perschy

Seven hundreds
Call Northside 777...............(1948) James Stewart, Richard Conte, Lee J. Cobb

Nine hundreds
976-Evil ....(1988) Stephen Geoffreys, Patrick O'Bryan, Sandy Dennis

Thousands
House of 1000 Corpses.................... (2003) Sid Haig, Karen Black, Bill Moseley
A Thousand Acres............... (1997) Michelle Pfeiffer, Jessica Lange
A Thousand Clowns .... (1965) Jason Robards, Barbara Harris, Martin Balsam
THX 1138 .... (1971) Robert Duvall, Donald Pleasence, Don Pedro Colley
1408 ...... (2007) John Cusack, Samuel L. Jackson, Mary McCormack
1492: Conquest of Paradise..................... (1992) Gérard Depardieu, Sigourney Weaver

Murder at 1600..... (1997) Wesley Snipes, Diane Lane, Daniel Benzali
1776........(1972) William Daniels, Howard Da Silva, Ken Howard
1900........................... (1976) Robert De Niro, Gérard Depardieu, Dominique Sanda
The Legend of 1900............(1998) Tim Roth, Pruitt Taylor Vince, Bill Nunn
1941 .... (1979) John Belushi, Dan Aykroyd, Treat Williams
1984..........(1984) John Hurt, Richard Burton, Suzanna Hamilton
1991: The Year Punk Broke ....(1992) Sonic Youth, Kim Gordon, Lee Ranaldo

Two thousands
Cherry 2000 .... (1987) Melanie Griffith, David Andrews, Pamela Gidley
Dracula 2000.....................(2000) Gerard Butler, Justine Waddell, Jonny Lee Miller
Death Race 2000......... (1975) David Carradine, Sylvester Stallone
2001 Maniacs ....(2005) Robert Englund, Lin Shaye, Giuseppe Andrews
2001: A Space Odyssey ....(1968) Keir Dullea, Gary Lockwood
2001: A Space Travesty ...... (2000) Leslie Nielsen, Ophélie Winter, Ezio Greggio
Blade Runner 2049...............(2017) Harrison Ford, Ryan Gosling, Ana de Armas
2010................. (1984) Roy Scheider, John Lithgow, Helen Mirren
2012........(2009) John Cusack, Thandie Newton, Chiwetel Ejiofor
2046 .... (2004) Tony Chiu-Wai Leung, Ziyi Zhang, Faye Wong
2047: Sights of Death .... (2014) Danny Glover, Daryl Hannah
Cyborg 2087.....(1966) Michael Rennie, Karen Steele, Wendell Corey
2307: Winter's Dream...............(2016) Paul Sidhu, Branden Coles, Arielle Holmes

Three thousands
Mystery Science Theater 3000: The Movie........ (1988) Joel Hodgson
Mr 3000............(2004) Bernie Mac, Angela Bassett, Michael Rispoli

3000 Miles to Graceland .... (2001) Kurt Russell, Kevin Costner, Courteney Cox

Tens of thousands
The 5,000 Finger of Dr. T......... (1953) Peter Lind Hayes, Mary Healy
10,000 BC.............. (2008) Camilla Belle, Steven Strait, Marco Khan
20,000 Leagues Under the Sea...... (1954) Kirk Douglas, James Mason

Millions
The $1,000,000 Reward ...... (1920) Lillian Walker, Coit Albertson
One Million Years BC ..... (1966) Raquel Welch, John Richardson, Percy Herbert
Million Dollar Baby........... (2004) Hilary Swank, Clint Eastwood, Morgan Freeman
Millions .... (1991) Billy Zane, Lauren Hutton, Carol Alt
Millions .... (2004) Alex Etel, James Nesbitt, Daisy Donovan
Brewster's Millions................... (1984) Richard Pryor, John Candy, Lonette McKee
20 Million Miles to Earth...... (1957) William Hopper, Joan Taylor
Infinity..... (1996) Matthew Broderick, Patricia Arquette, Jeffrey Force

Times
3.10 to Yuma..... (2007) Russell Crowe, Christian Bale, Ben Foster
3.10 to Yuma.............. (1957) Glenn Ford, Van Heflin, Felicia Farr
Four in the Morning........................ (1965 Ann Lynn, Judi Dench, Norman Rodway
11:14 .... (2003) Henry Thomas, Colin Hanks, Ben Foster
10 to Midnight .... (1983) Charles Bronson, Lisa Eilbacher, Andrew Stevens

APPENDIX C

BOOKS WITH NUMBERS IN THEIR TITLES

Zero
And Then There Were None by Agatha Christie

One
The One That Got Away by Chris Ryan
One Flew Over the Cuckoo's Nest by Ken Kesey
One Fish, Two Fish, Red Fish, Blue Fish by Dr Seuss
(One, Two etc…) for the Money (series) by Janet Evanovitch
The No. 1 Ladies Detective Agency by Alexander McCall Smith
One Day in the Life of Ivan Denisovich by Aleksandr Sozhenitsyn

Two
The Two Towers by J R R Tolkien
Bravo Two Zero by Andy McNab
The Second Coming by W B Yeats
At Swim–Two Birds by Flann O'Brien
A Tale of Two Cities by Charles Dickens
Sybil or the Two Nations by Benjamin Disraeli

Three
Life of Pi by Yann Martel
Three Lives by Gertrude Stein
The Three Sisters by May Sinclair
The Third Man by Graham Greene
Three Soldiers by John Dos Passos
Three Act Tragedy by Agatha Christie
Three Men in a Boat by Jerome K Jerome
The Three Musketeers by Alexandre Dumas
The Three Paradoxes by Paul Hornschemier
The Drawing of the Three by Stephen King
Three men on the Bummel by Jerome K Jerome
Wild Swans: Three Daughter of China by Jung Chang

Four
The Four Loves by C S Lewis
The Big Four by Agatha Christie
The Four Just Men by Edgar Wallace
The Sign of the Four by Arthur Conan Doyle

Five
High Five by Janet Evanovich
Five Children and It by E Nesbit
Slaughterhouse 5 by Kurt Vonnegut
The Fifth Elephant by Terry Pratchett
Five Red Herrings by Dorothy L Sayers
The Famous Five (series) by Enid Blyton
The Five People You Meet in Heaven by Mitch Alborn

Six
Six Years by Harlan Coben
Six of One by Rita Mae Brown
Six Crows by Leigh Bardugo
The Secret Seven (series) by Enid Blyton

Seven
Seven Troop by Andy McNab
The Seven Voyages of Sinbad
Seven Sisters by Earlene Fowler
Seven Years in Tibet by Heinrich Harrer
The Secret Seven (series) by Enid Blyton
Seven Types of Ambiguity by Elliot Perlman
The Seven Dials Mystery by Agatha Christie

Eight
The Eight by Katherine Neville
Eight Cousins by Louisa M Alcott

Nine
Nine Stories by J D Salinger
Nine Tomorrows by Isaac Asimov
Nine Lives to Die by Rita Mae Brown
The Nine Taylors by Dorothy L Sayers

Ten
10lb Penalty by Dick Francis
Starter for Ten by David Nicholls
Ten Apples Up on Top by Dr Seuss
Ten Little Ladybugs by Melanie Gerth
A History of the World in 10½ Chapters by Julian Barnes

Eleven
Eleven by Highsmith
Eleven Birthdays by Wendy Mass

Twelve
Twelve Angry Men by Reginald Rose
Twelve Red Herrings by Jeffrey Archer
Twelve Years a Slave by Solomon Northup

Thirteen
Thirteen Reasons Why by Jay Asher
The Secret Diary of Adrian Mole, Aged 13¾ by Sue Townsend

Fourteen
Fourteen by Peter Clines

Fifteen
Fifteen Minutes by Karen Kingsbury
Dick Sand, or: A Captain at Fifteen by Jules Verne

Sixteen
Sixteen Cows by Lisa Wheeler

Seventeen
At Seventeen by Gerri Hill

Eighteen
Eighteen Summers by Jessie M

Nineteen
Nineteen Minutes by Jodi Picoult

Twenties
Catch-22 by Joseph Heller
Twenty-Five by Rachel L. Hamm
Twenty-Six Roses by Tamara Vann
Twenty-Three Tales by Leo Tolstoy
Twenty Years After by Alexandre Dumas
Twenty-Seven Bones by Jonathan Nasaw
The Twenty-One Balloons by William Pene du Bois
Twenty Thousand Leagues Under the Sea by Jules Verne
Twenty-Eight and a Half Wishes by Denise Grover Swank
The Rabbi and the Twenty-Nine Witches by Marilyn Hirsh
The Twenty-Four Days Before Christmas by Madeleine L'Engle
The Twenty-Three Days of the City of Alba: Stories by Fenolgio

Thirties
31 Songs by Nick Hornby
Thirty-Seven by Maria Beaumont
Thirty-Six Hours by Allison Brennan
Thirty-Three Teeth by Colin Cotterill
The 34th Degree by Thomas Greanias
Thirty-Eight Nooses by Scott W. Berg
The Thirty-Nine Steps by John Buchan
Thirty Seconds by Heather MacPherson
Thirty-Two Going on Spinster by Becky Monson
Stranger on Route Thirty-Five by Leslie Sansom
Thirty-One Dates in Thirty-One Days by Tamara Duricka Johnson

Forties
Forty-Four by Jools Sinclair
Forty-Five by Bill Drummond
Forty-Six Pages by Scott Liell
Forty-Eight X by Barry Pollack
Ali Baba and the Forty Thieves
North Dallas Forty by Peter Gent
48 Shades of Brown by Nick Earls
Forty-Two by M. Thomas Cooper
Forty-Seven Kisses by Victoria Grant
The Forty Rules of Love by Elif Shafak
The Crying of Lot 49 by Thomas Pynchon
The Crying of Lot 49 by Thomas Pynchon
Forty-One Jane Doe's by Carrie Olivia Adams
Forty-Three Septembers by Jewelle Gomez
The Forty-Five Guardsmen by Alexandre Dumas
Forty-Six Years in the Army by John McAllister Schofield

Fifties
Fifty-Four by Wu Ming
Fifty-Six Men by Fred Placke
Fifty Three by Rosanita Ratcliff
Fifty Shades Freed by E L James
Fifty Shades Darker by E L James

Fifty Shades of Grey by E L James
Fifty-Nine in '84 by Edward Achorn
Fifty-Two Pickup by Elmore Leonard
Fifty-Five Graves by Robert P. Maroney
Fifty-One Tales by Edward J.M.D. Plunkett
Fifty-Two Days by Camel by Lawrie Raskin
The Fifty Year Sword by Mark Z. Danielewski
The Folks at Fifty-Eight by Michael Patrick Clark
Grey: Fifty Shades as Told by Christian by E L James
Darker: Fifty Shades as Told by Christian by E L James
The Fifty-Seven Lives of Alex Wayfare by M.G. Buehrlen

Sixties
Highway 61 by William McKeen
Sixty-Six Chances by Nick Buxton
The Sixty-Four Sonnets by John Keats
Highway 62 Revisited by Matt Maxwell
The Sixty-Eight Rooms by Marianne Malone
The House of Sixty Fathers by Meindert DeJong
Sixty-Nine Inches and Rising by Rebecca Steinbeck
The Sixty-Five Years of Washington by Juan Jose Saer
Scandinavian September Sixty-Nine by Bennett Lear Fairorth
Sixty-Seven Poems for Downtrodden Saints by Jack Micheline

Seventies
Seventy-Eight Days by Ciara Howard
Seventy-Three Poems by e.e. cummings
Seventy-Nine Park Avenue by Harold Robbins
The Book of Seventy by Alicia Suskin Ostriker
Seventy-Six Days on Mars by Michael Redman
Seventy-Seven Shadow Street by Dean Koontz
Seventy-Two Hour Hold by Bebe Moore Campbell
Seventy-Eight Degrees of Wisdom by Rachel Pollack
Seventy-Four Seaside Avenue by Debbie Macomber
Seventy-Five Pretzels: A New York Tale by Michael Marzi
Seventy-One Days: The Media Assault on Obama by Michael Jason Overstreet

Eighties
GB84 by David Peace
Eighty-Eight by A.L. McAuley
Eighty-Two Desire by Julie Smith
The 89th Kitten by Eleanor Nilsson
Eighty-Nine Pounds by Lauren Groff
84, Charing Cross Road by Helene Hanff
The Ghosts on 87th Lane by M.L. Woelm
Around the World in 80 Days by Jules Verne
Eighty-One Miles: Best Loved Poems by Shane Windham
Eighty-Five Days: The Last Campaign of Robert Kennedy by Jules Witcover

Nineties
Ninety-Three by Victor Hugo
Ninety-One Days by Lindsay Luterman
The Ninety and Nine by Vinny DiGirolamo
God Has Ninety-Nine Names by Judith Miller
Ninety-Two in the Shade byThomas McGuane

Hundreds
The Hundred and One Dalmatians by Dodie Smith
100 Selected Poems by E. E. Cummings by E. E. Cummings
One Hundred Years of Solitude by Gabriel Garcia Marquez

Three hundreds
Andy Warhol 365 Takes: The Andy Warhol Museum Collection by Staff of Andy Warhol Museum

Four hundreds
Fahrenheit 451 by Ray Bradbury

Five hundreds
The 500 Hats of Bartholomew Cubb by Dr. Seuss

Thousands
1985 by Anthony Burgess
1408 by Stephen King
Nineteen Eighty-Four - Orwell
A thousand Acres by Jane Smiley
One Thousand and One Nights (stories)
Nineteen Seventy Four by David Peace
Nineteen Seventy Seven by David Peace
2001: A Space Odyssey by Arthur C Clarke
A Thousand Splendid Suns by Khaled Hosseini
1,000 Places to See Before You Die by Patricia Schultz
1001 Movies You Must See Before You Die by Steven Jay Schneider (Editor)

Tens of thousands
20,000 Leagues under the Sea by Jules Verne
Illustrated 10,000 Dreams Interpreted: An Illustrated Guide to Unlocking the Secrets of Your Dreamlife - Gustavus Hindman Miller

Millions
A Million Little Pieces by James Frey
Brewster's Millions by Richard Greaves

Dates and times
11/22/63 by Stephen King
Four Past Midnight by Stephen King
4:50 from Paddington by Agatha Christie

APPENDIX D

NUMBERS WHICH OCCUR IN THE BIBLE

There are sections of the Bible which feature numbers quite prominently including the Book of Numbers itself. It is reputed that the number forty occurs 126 times and seven is very common. The paragraphs which follow list some of the most important ones based on the KJV.

The Old Testament

Genesis
5: Generations after Adam, Methuselah at 969.
6: Dimensions of Noah's Ark
7 and 8: Loading the ark and after the flood
9: The age of Noah, 950 years at his death
11: More generations
18: Fifty righteous in Sodom and Gomorrah
40 and 41: Joseph's dream interpretations

Exodus
7 to 11: The Plagues of Egypt:

First	Waters to blood	**Second**	Frogs
Third	Lice	**Fourth**	Flies
Fifth	Death of cattle	**Sixth**	Boils
Seventh	Thunder and hail	**Eighth**	Locusts
Ninth	Darkness	**Tenth**	Death of firstborn

12 and 13: The Passover
14 and 15: Pharaoh's pursuit
20: The Ten Commandments
25 and 37: Design and making of the Ark of the Covenant
26 and 36: Design and making of The Tabernacle
27: Design of The Court of The Tabernacle
30 and 38: Design and making of The Altar
31 and 35: The Sabbath
39: Making of The Breastplate

Numbers
1 and 2: number the army of each tribe of Israel and their battle order:

Tribe	Number	
Reuben	46,500	
Simeon	59,300	set forth second
Gad	45,650	151,450
Judah	74,600	
Issachar	54,400	set forth first
Zebulon	57,400	186,400
Ephraim	40,500	
Manasseh	32,200	set forth third
Benjamin	35,400	108,100
Dan	62,700	
Asher	41,500	set forth hindmost
Naphtali	53,400	157,600
	603,550	

(The Levites do not contribute as they manage the Tabernacle)

Deuteronomy
9: Moses 40 days and 40 nights on the mount and two tablets of stone

Joshua
4: Twelve men, twelve stones and the waters of Jordan
6: The fall of Jericho
21: The cities of the Levites

Judges
10: Thirty sons, ass colts and cities
12: Thirty sons, daughters and nephews
14: Thirty sheets and changes of garments
17: Eleven hundred shekels etc…
20: Battle numbers at Gilbeah

1 Samuel
13: Saul smites the Philistines
17: David and Goliath

1 Kings
4: Solomon's provisions
5: Solomon's levy
6 and 7: The dimensions of Solomon's Temple
10: Solomon's throne and wealth
11: Solomon's wives and concubines

2 Kings
7: Four leprous men, two measures of barley for a shekel
11: Hundreds and Jehoash king at seven

1 Chronicles
7: Issachar's descendents
12: David's armed men
15: The children of Aaron and the Levites
18: David smites Hadarezer
21: Joab gives numbers to David
23 – 27: Divisions and sons

2 Chronicles
3: Solomon and the house of God
4: Solomon and the altar
9: Solomon's wealth and throne

Ezra
2: Numbers who returned from Babylon
Nehemiah
7: Numbers of those returning from Babylon
11: Those who went to live in Jerusalem

Job
1: Job's family and stock
Jeremiah
52: Zedekiah, Nebuchadnezzar and captives

Ezekiel
40: Dimensions of the house, wall and court
41: Dimensions of the temple
42: Dimensions of the courts and chambers
43: Dimensions of the altar
45: Dimensions of the sanctuary
48: Dimensions of the city

Daniel
3: The fiery furnace
7: The fourth beast

The New Testament

Matthew
4: Forty days and forty nights in the wilderness
14: Five loaves and two fishes to feed the 5,000; 12 baskets of leftovers
15: Seven loaves, a few small fishes to for 4,000; 7 baskets of leftovers
16: Feeding the 5,000 and the 4,000 recalled

22: One bride for seven brothers
25: The foolish virgins and every man according to his ability
26: Judas and 30 pieces of silver plus Peter's thrice denial

Mark
1: Forty days in the wilderness
4: Parable of the seed
6: Five loaves and two fish to feed the 5,000; 12 baskets of leftovers
15: Seven loaves, a few small fishes to for 4,000; 7 baskets of leftovers
12: Seven brothers for one bride
14: Peter denies Jesus

Luke
4: Forty days in the wilderness
9: Five loaves and two fish to feed the 5,000; 12 baskets of leftovers
20: Seven brothers for one bride
22: Peter denies Jesus

John
6: Five barley loaves and two fish to feed the 5,000; 12 baskets of leftovers
13/18: Peter denies Jesus

Revelation
1: Seven stars and seven candlesticks
4: Four and twenty elders and beasts etc.
5: The voice of many angels
6: The third and fourth seals
7: The four corners of the earth
8: The seventh seal
9: The fifth angel
11: The seventh seal
12: A great red dragon
13: A beast with seven heads and ten horns
15: Seven angels having seven last plagues
16: All seven angels

17: Seven angels
19: Four and twenty elders and four beasts
20: A thousand years
21: Twelve gates, twelve angels and the city
22: Twelve manner of fruits